Advances in Intelligent and Soft Computing 121

Editor-in-Chief: J. Kacprzyk

T0180111

Advances in Intelligent and Soft Computing

Editor-in-Chief

Prof. Janusz Kacprzyk
Systems Research Institute
Polish Academy of Sciences
ul. Newelska 6
01-447 Warsaw
Poland
E-mail: kacprzyk@ibspan.waw.pl

Further volumes of this series can be found on our homepage: springer.com

Mariusz Kaleta and Tomasz Traczyk (Eds.)

Modeling Multi-commodity Trade: Information Exchange Methods

 Springer

Editors
Mariusz Kaleta
Warsaw University of Technology
Institute of Control and Computation
  Engineering
Warsaw
Poland

Tomasz Traczyk
Warsaw University of Technology
Institute of Control and Computation
  Engineering
Warsaw
Poland

ISSN 1867-5662
ISBN 978-3-642-25648-6
DOI 10.1007/978-3-642-25649-3
Springer Heidelberg New York Dordrecht London

e-ISSN 1867-5670
e-ISBN 978-3-642-25649-3

Library of Congress Control Number: 2011946100

Printed on acid-free paper

Springer is part of Springer Science+Business Media (www.springer.com)

Preface

Probably most of us are conscious witnesses of market mechanism entrance into new areas of human activities in a real life. Everywhere, the distribution and allocation of goods take place, temptations and incentives for market rules introduction appear. But not all areas of human activities are identical and follow the same principles. Thus, simple transfer of a market solution between fields of activity cannot guarantee good quality of solution and meeting all expectations.

In parallel to real life, researches of market mechanisms are being conducted since many years, including many different fields of science. One, important field is mechanism design theory, sometimes called reverse game theory. However, despite huge achievements, it appears that there are still not satisfactory tools and methodologies. It is especially clear if we confront the state of the art with new challenges coming from new areas of market mechanism applications. In these new circumstances, many new additional aspects must be addressed.

One of the important aspects is a trade *under constraints*. Market mechanisms are entering into new fields of economy, in which trade world must take into account physical world. Physical aspects influence the possibilities for transport and exchange of the market commodities, hence usually they can be modeled by a graph representing infrastructure. For instance, the graph can represent power grid in electrical energy system, telecommunication infrastructure, or network of roads. In each of these cases, there is some set of limited resources. Whole community of market participants uses the infrastructure and needs related resources in order to organize goods and services exchange.

Another specific issue is a *real-time*. Under permanent pressure of competitors, market world tends to react upon stimulus at shortest possible time. This tendency brings many new notions and ideas, which essence lies in notion of real-time enterprize. Under strong time requirements and permanent lack of time, the idea to delegate some competency to autonomous software components, that is agents, is very attractive. But not only pressure of competition forces to turn into real-time aspects. Also, some of market commodities are not storable. For instance, telecommunication bandwidth is available at a given time slot and if it is not sold it generates lost profits. We do not have effective methods for storing the electrical energy in a system scale as well.

Next, new aspects come out from *market entities requirements*. In the case of infrastructure and real-time requirements, expression of entities preferences by simple offers is not satisfactory. Also, mechanisms known, for instance, from combinatorial auctions do not meet all functional expectations.

Multi-commodity is a derivative of previously mentioned aspects. Infrastructural aspects, functional and real-time requirements make incentive for multi-commodity solutions. In complex market mechanisms many commodities emerge. Some commodities can be related to transmission services in infrastructure net or commodities conversion. Some of them can be related to time structure – if some type of commodity located in different time slots becomes two commodities. Rich commodity structure can serve building rich functionality for market participants and helping them to express their preferences. But not only richness of commodities structure is important. The key is to be able to model and consider relationships between the commodities. Multi-commodity mechanism does it. It allows to formulate new, more sophisticated models and solutions, which reflect decision situations in a better manner.

Application of multi-commodity approach in practice requires solving several issues. Exemplary issues are related to data modeling in information systems, communication models between market participants, semantics aspects of communication, reliability, and others. This book answers some of the questions and points out most promising and attractive paths for implementation and development. In the center of the book we put our data and communication model for multi-commodity markets, that is M^3 (Multi-commodity Market Model). M^3 is a set of open models which facilitate instantiations of multi-commodity markets. Since expressiveness is one of the main goals of M^3, rich variety of possible market designs can modeled. Obviously, both the M^3 model and this book do not provide complete solution for all issues related to market trade modeling. In the book we raise information issues from system, model point of view.

This book contains revised versions of papers presented during scientific workshop "Modeling Multi-commodity Trade: Information exchange methods", which took place in November 2010 at Warsaw University of Technology. It summarizes results of the research work supported so far by scientific grant "Methods and architectures of information interchange for electronic trade on infrastructural markets" (see page IX), and some earlier research work on multi-commodity markets modeling. Though partial results of the research were published earlier, the book gives the most complete view on results of our research in the field of modeling the trade on complex multi-commodity infrastructural markets.

The book is divided into two parts. The first part starts with the basics of Multi-commodity Market Model (M^3): our motivation to create the model is shown, and theoretical foundations of the model are introduced. Next, the M^3 data model is presented, along with its implementation in XML. Subsequent chapters focus on communication and architectural issues, with special interest in multiagent systems. Last two chapters of the first part discuss supplementary matters: possibilities to use ontologies in the context of M^3 and reliability aspects of multi-commodity trade.

The second part of the book discusses possible applications of M^3 in various complex multi-commodity markets. First, applications of M^3 in electricity market are shown; this market is very characteristic for multi-commodity trade in infrastructure networks, and complex problems of this market induced us to create M^3 model. Next, issues of pollution emission permits trade are presented and opportunities to use M^3 in this growing market are shown. Last, possible applications of M^3 in telecommunications, precisely in virtual network market, are described.

Warsaw,   Mariusz Kaleta
September 2011   Tomasz Traczyk

Acknowledgements

The research was supported by the Polish National Budget Funds 2009-2011 for science under the grant N N516 3757 36.

Contents

Part II: M^3 Applications

List of Contributors

Tomasz Gidlewski
Warsaw University of Technology, Institute of Control and Computation
Engineering
e-mail: T.Gidlewski@stud.elka.pw.edu.pl

Janusz Granat
Warsaw University of Technology, Institute of Control and Computation
Engineering
e-mail: J.Granat@ia.pw.edu.pl

Joanna Horabik
Polish Academy of Sciences, System Research Institute
e-mail: Joanna.Horabik@ibspan.waw.pl

Przemysław Kacprzak
Warsaw University of Technology, Institute of Control and Computation
Engineering
e-mail: P.Kacprzak@elka.pw.edu.pl

Mariusz Kaleta
Warsaw University of Technology, Institute of Control and Computation
Engineering
e-mail: M.Kaleta@ia.pw.edu.pl

Kamil Kołtyś
Warsaw University of Technology, Institute of Control and Computation
Engineering
e-mail: K.J.Koltys@elka.pw.edu.pl

Jacek Malinowski
Polish Academy of Sciences, System Research Institute
e-mail: Jacek.Malinowski@ibspan.waw.pl

Zbigniew Nahorski
Polish Academy of Sciences, System Research Institute
e-mail: Zbigniew.Nahorski@ibspan.waw.pl

Piotr Pałka
Warsaw University of Technology, Institute of Control and Computation
Engineering
e-mail: P.Palka@ia.pw.edu.pl

Krzysztof Pieńkosz
Warsaw University of Technology, Institute of Control and Computation
Engineering
e-mail: K.Pienkosz@ia.pw.edu.pl

Weronika Radziszewska
Polish Academy of Sciences, System Research Institute
e-mail: Weronika.Radziszewska@ibspan.waw.pl

Dominik Ryżko
Warsaw University of Technology, Institute of Computer Science
e-mail: D.Ryzko@ii.pw.edu.pl

Kamil Smolira
Warsaw University of Technology, Institute of Control and Computation
Engineering
e-mail: K.Smolira@ia.pw.edu.pl

Jarosław Stańczak
Polish Academy of Sciences, System Research Institute
e-mail: Jaroslaw.Stanczak@ibspan.waw.pl

Eugeniusz Toczyłowski
Warsaw University of Technology, Institute of Control and Computation
Engineering
e-mail: E.Toczylowski@ia.pw.edu.pl

Tomasz Traczyk
Warsaw University of Technology, Institute of Control and Computation
Engineering
e-mail: T.Traczyk@ia.pw.edu.pl

Przemysław Więch
Warsaw University of Technology, Institute of Computer Science
e-mail: P.Wiech@ii.pw.edu.pl

Part I
Multi-commodity Market Model (M^3) Fundamentals

M^3 – Motivations and Formal Model

Mariusz Kaleta and Eugeniusz Toczyłowski

Abstract. In the chapter, first we discuss needs and motivations for market-based mechanism for complex, distributed systems, especially focusing on infrastructures sectors. A great number of issues related to market mechanisms developing is modeled by decisions models. There are many completely different attempts to organizing the mechanisms of finding market equilibrium. These attempts are hard to reliably compere on the common test cases due to huge effort needed for numerical data adaptation to each approach. We give strong foundations for Multicommodity Market Model (M^3), which describes the data on the input and output sides of the market clearing process. Its main advantage is very high expressiveness. M^3 covers most known trade mechanisms, including different auction types, multi-commodity mechanisms, bilateral contracts and others. In the chapter we present architecture of M^3, including formal model.

1 Introduction

Complex systems are often beyond efficient direct control and management. Thus, during the past decades, the world-wide market liberalization and deregulation processes are being implemented in many network infrastructure sectors, including power systems, telecommunication, computer, rail and transport networks, water, urban systems and others.

Under deregulation, the systems are undergoing drastic restructuring and transformation from cost-conscious, regulated utilities to competitive market participants. These entities have their own independent interests, values, different tasks,

Mariusz Kaleta
Warsaw University of Technology, Institute of Control and Computation Engineering
e-mail: M.Kaleta@ia.pw.edu.pl

Eugeniusz Toczyłowski
Warsaw University of Technology, Institute of Control and Computation Engineering
e-mail: E.Toczylowski@ia.pw.edu.pl

M. Kaleta & T. Traczyk (Eds.): Modeling Multi-commodity Trade, AISC 121, pp. 3–19.
springerlink.com   © Springer-Verlag Berlin Heidelberg 2012

operations and services. State-owned or private monopolies, that have been functioned traditionally in the infrastructure sectors, are being gradually transformed into various market entities, which must operate in new competitive market environments, under regulated market rules, with operational help of various market institutions, such as auctions, commodity exchanges, real-time balancing markets, etc.

In the deregulated framework for control and management, instead of operating according to central rules and plans established by a hierarchical control structure in a centralized system, the systems operate through cooperative behavior of many entities which interact as the competitive market participants. The need for better management and control of large distributed network systems stimulated in research community a great deal of interest in developing new competitive market mechanisms for management and closed-loop operational control procedures to help system performance optimization. An important stream of the research work on market development is focussed on gradual functional decentralization with allocation of obligations and rights to the distributed market-players. On another side, some market integration processes are also enhanced.

One difficulty is in analyzing and comparing different market models, methods, experiments and solutions in a reliable way, due to heterogenous experiment environments, different data models, and various numerical data and data storage methods that are used. However, the market models that are different at a first glance, in fact operate on the same types of data. Moreover, only the best auction clearing mechanisms verified by many researches in multiple experiments should be implemented as practical solutions. Therefore, from perspective of serving the development and verification of new auction mechanisms, the open market information interchange systems, including data models, are extremely important for the future of the network industries. On the real markets, the tendencies to operate in a real-time and to react rapidly to the demand side requirements, strengthen even more the need for standardization models [4].

The market processes consist of a sequence of many elementary balancing and clearing processes that tend towards a complete system's balance at a real time. Usually, each process has its own mechanism for information interchange and processing. At present, there are no general world-wide standards for information interchange mechanisms. In some industries there are data interchange mechanism which can be acknowledged as local standards, for instance, RosettaNet standards in electronic industry, MDDL (Market Data Definition Language) in financial sector and other industries standards specified on the basis of open standards like ebXML (Electronic Business using eXtensible Markup Language) or XBRL (eXtensible Business Reporting Language) [6, 7, 8, 9]. However, these standards are focused on electronic communication of business and financial data like invoices, offers, business partner information and so on. Electrical energy markets are one of the most advanced and complex infrastructure markets, however also here some standards are developed. In energy sector, UCTE (Union for the Co-ordination of Transmission of Electricity) initiative called Electronic Highway focuses only on transport layer and technical aspects of the communication network. It is clear that existing standards cannot meet the needs of the infrastructural sector, where many specific elements, related to real-time balancing of many commodities and services under constraints, may play important roles. Heterogenous systems and absence of general mechanism for data interchanges create barriers for mechanism integration and developments that are especially important for the European and world-wide market evolutions.

Market entities create, process, and consume extensive volumes of market data and must deal with dizzying collection of sources, necessary to conduct market operations. Data users need to devote considerable resources in translating data from multiple sources. In order to integrate various market solutions and to ease mappings between various market data applications and systems, an open Multi-commodity Market Model (abbr. M^3) was proposed in [2, 3]. M^3 is a formal data information model that may be used in designing open information systems for bidding, balancing and clearing of the market processes, in the context of multi-commodity trade in various network infrastructure sectors. One advantage of using the M^3 model is an easy exchange of all data between various market entities and complex market balancing processes, within the XML-derived information interchange specification, based on proposed M3-XML dialect.

A set of definitions of M^3 promotes clear understanding and standard interfaces for the data flow. The common market data terminology embraced by M^3 allows one to clearly state the nature and origin of the structured market data elements, thus removing ambiguity. The particular advantage of M^3 is that it may help the designers to conduct some development simultaneously and to some extent independently, whereas, thanks to a unified model, data and results may be easily exchanged or/and shared between the users of various market clearing systems. M^3 may be useful in facilitating communication and coordination between subsequent market sub-areas and stages. It may be very helpful both for the market operators as well as for the market users, as it facilitates data exchange procedures between the market operators and market users. It may also be crucial for achieving the long-term benefits of interoperability and enhancement of market integration and quality of market products.

An important goal of M^3 is also to integrate market data from multiple distributed sources in diverse systems throughout the global "enterprise" network without having to understand how information providers format and normalize internal processes and data. The payoff is to shift the focus of internal efforts from issues associated with the "formatting of data" to those associated with the "quality of processes" to enhance the market functionality.

M^3 information model is helpful in designing a distributed information environment for bilateral or multilateral multi-commodity trade in the multi-agent systems. In that environment a variety of participants can operate and compete; each of them has either active or passive role in the context of communicating with other participants. Passive participants submit offers and wait for acceptance. Active participants seek offers available in the network and try to earn maximum profits either from mediation between some other participant's offers, or from negotiating and buying /selling goods. Some active participants may also act as the (profit-neutral) system operators. The distributed multi-agent market environment can be an alternative for the existing types of centrally organized auctions.

2 Market Design

At present, in many network industries, functionalities of the existing market designs are not completely satisfactory. For a complete successful market system design, reactions to all operational situations should result from market-driven processes for achieving economic market equilibria, together with considering technical and operational conditions, resource constraints and/or automatic control system requirements. Due to operational and real-time requirements existing in such systems, specific problem-oriented market designs are usually required.

Many researches and professionalists around the world participate in development, investigation and implementation of a variety of new ideas related to auction and market clearing systems under various market conditions. At present, there are no conformity in the research directions that can be considered as the most promising ones. Despite a tremendous world-wide research it seems that the decision makers still do not possess enough knowledge to direct the market evolution and support the best directions in which the market systems should evolve.

In the network systems, an efficient market balance may be obtained in a single balancing process by joint optimization of trade of many elementary commodities and services related to buy and sell offers of the network resources. For this purpose the multi-commodity exchanges can be used, in addition to single-commodity exchanges and bilateral trading. The basic multi-commodity market clearing model is in the LP form and enables maximizing global economic welfare and effective balancing of sell and buy offers for bundles of elementary commodities [5]. It has all positive features of the classical single-commodity market clearing model, yet enabling handling many real-world requirements.

Apart from traditional auctions, long-term and medium-term single-commodity market segments, or day-ahead and intra-day-markets, there is a need for designing specific problem-oriented multi-commodity auctions and balancing market mechanisms, which must provide feasible execution of sales contracts and assure timely delivery of many goods and services.

In the case of infrastructure sectors, implementing free market trading appears to be significantly harder then in the other sectors. Trading and deliveries of goods and services usually needs some limited resources. Existence of various resource constraints strongly contributes to some costs and occurrence of the local market power. Resource limitations reduce the freedom of the market competitive solutions and may decrease profits, i.e, values of the economic wealth that would result from the free market trade. Moreover, other more technical and security aspects, that influence the market solutions, must be taken into account. A more general multi-commodity market clearing model developed in [5] is in the MILP form. Multiple clearing on forward and real-time market segments assures that the supply and demand for many goods and services can be matched simultaneously in real time under various constraints and requirements.

Achieving efficient market equilibria in constrained network systems in real time is a challenging task for researchers. Various market segments must be designed for ensuring safe, feasible and economically efficient system balances, that can be achieved in real time in all possible operational situations. The forward and real-time balancing markets should also allow the market players to change their preferences and to modify correspondingly the offers and bids, in a short time just before delivery. In this way the market players may react to rapid changes of the system state. Apparently, also the closed-loop feedback control system must be harmonized with the real-time market processes.

In this book we address appropriate market design issues and control mechanisms for distributed networks. One purpose of the open M³ model is that it creates a flexible framework for development of new market models and algorithms, benchmark data repository, and gives possibility for integration of software components which implement balancing mechanisms. Finally, it will help the community to determine the best industrial standards of data interchange and enable for an easy public access and exchange of various market data.

3 Market Game

We consider a distributed environment of many agents. The agents interact in order to exchange goods and services, both called the commodities here and after. A market system constitutes the rules for agents interactions in order to solve a given balancing problem (BP). A mediator (market maker) is an agent who collects offers and specifies accepted volumes and market prices. An extreme arrangement of the market system is a fully distributed one. In this kind of market system there is no central or quasi-central (locally central) mediator. Agents just negotiate bilaterally or multilaterally. On another side, the market system can be centralized in the meaning of one central mediator as a sole way to contract. Since mediator usually is supposed to take the best decisions for agents community, such systems are referred as optimized. The mediator runs a mechanism responsible for market operation. The mechanism consists in decision models in which commodities allocations and payments are determined to achieve a given market objectives.

3.1 Mechanism

Mechanism can be viewed as a game, for which desired result is specified by a function F. Let us assume that I agents participate in this game. Each of them has its own preferences, usually not known to market designer. More over, the preferences are usually against a desired result specified by function F. However, a mechanism *implements* function F if observed game solution states an equilibrium point and this equilibrium converges to result given by function F [1]. Desired result is usually specified by a set of properties, that should be satisfied. Typically, desired properties are not compatible with agent preferences.

A market mechanism gives accepted volumes of transactions, so called quantitative solution, and it gives settlement prices, so called value solution. Quantitative balancing results from demand and supply volumes. Usually equilibrium of demand and supply volumes is achieved as a solution. Value solution is related to individual payments of each agent.

An environment of a mechanism is a tuple $(N, \Theta, \mathscr{A}, \mathscr{U})$, where N is an agents set, Θ is a set of market states. Among others things, a market state includes agents offers and preferences. As a result of mechanism execution, quantitative results $a = a(\Theta) \in \mathscr{A}$ are achieved. Value results a consist in market parameters $x = x(\Theta) \in \mathscr{X}$, e.g. market prices, offers attributes, and quantitative results $o = o(\Theta) \in \mathscr{O}$, mainly turn volumes. As a result, the individual value outcomes are obtained, $u_i(a, \Theta) \in \mathscr{U}_i, i \in N$. They depends on parameters x and quantitative results o. Set $\mathscr{A} \subseteq \mathscr{X} \times \mathscr{O}$ is a space of all quantitative results and set $\mathscr{U}$ is a space of all value results.

Let $f : \Theta \to \mathscr{A}$ be so called market choice function, which determined desired quantitative results of game $a = f(\Theta) \in \mathscr{A}$ for market state Θ. Agents take some strategies $s_i(\Theta_i) \in S_i, i \in N$, where S_i is a space of all potential strategies involved by market mechanism. A rule for market decision is a function $g : S_1 \times \ldots \times S_n \to \mathscr{A} \times \mathscr{U}$. Thus, market mechanism $\mathscr{M}$ is defined by pair (S, g).

For an environment $E = (N, \Theta, \mathscr{A}, \mathscr{U})$, mechanism $\mathscr{M} = (S, g)$, is an implementation of market choice function $f \to E$, if for each $\theta \in \Theta$, there is strategies vector $s^* = (s_1^*, \ldots, s_n^*)$ for which $g(s_1^*, \ldots, s_n^*) = f(\Theta)$ and s^* is an equilibrium point of game induced by mechanism $\mathscr{M} = (S, g)$.

A market mechanism embedded in an environment is presented in Figure 1. Agents, depicted as A1, A2, A3, SO in the figure, send their signals to the mechanism. There is one emphasized agent SO, who is called system operator or mediator and is responsible for mechanism execution. However SO also participates in market game. He sends some signals to the mechanism and can benefit from mechanism operation.

After collecting signals, mechanism is executed. Then signals are send back to agents, including allocations and prices determined by the mechanism. In the one-shot market, the allocation and prices are finally. In iterative case, after agents are enriched with back signals, they can again send new signals to the mechanism, until some stop conditions are not satisfied.

Even more complex situation is presented in the Figure 2. Here we assume, that agents need additional resources to provide commodities. The resources are modeled as a network. Each agent is located in a node of the network. Limited resources are modeled as arcs capacities or node divergences.

Market mechanism implementing market choice function can be evaluated in two areas. First is a quality of choice function, mainly general features, common for all agents and environment. Second, in a context of strategies and outcomes of each individuals, which are mostly individual aspects.

Properties of market mechanism $\mathscr{M}$ can be evaluated using the game theory tools. The most attractive mechanism are those with dominant equilibrium point. In this case optimal strategies $s_i^* \in S_i$ of agents are dominant strategies – no agent can benefit by changing is strategy, no matter how other agents will play.

In general, market choice function $f(\Theta)$ can be evaluated via vector of criteria $Q_1(a),\ldots,Q_m(a)$. Assuming that market state Θ is admissible, pareto-effective solutions can be searched. Market choice function is paret-optimal if for every market state, no criteria $Q_1,\ldots,Q_m$ can be improved without worsening other criteria.

3.2 Balancing Process

Mechanism implementation consists in many subprocesses which execute elementary tasks. There is a need for auxiliary elementary tasks like collecting signals from agents, results publishing and so on. The core of mechanism is performed in elementary balancing tasks (EBT), which realize allocation and valuation rules. It applies especially to iterated mechanisms, in which a sequence of elementary balancing tasks is constituted. Also in complex systems such structures are formed, e.g. balancing on energy market may be performed in a sequence of day-ahead, intra-day and real-time balancing markets. Moreover, in each mentioned market a sequence of elementary balancing tasks may exist. Despite the architecture of system may be quite complex, the elementary balancing tasks are essential, and whole architecture can be viewed as a composition of elementary balancing task supported by some auxiliary elementary tasks.

We treat an elementary balancing task as a "black box", which transforms some input data (market offers, network data, initial programmes, etc.) into output data: market prices, settlements, updated programmes, etc. We make no assumptions concerning the way how the input data can be processed. Thus, the task can be a pure auction or exchange mechanism, the real-time balancing, or any other balancing or clearing process. The elements of the open data model for the market systems

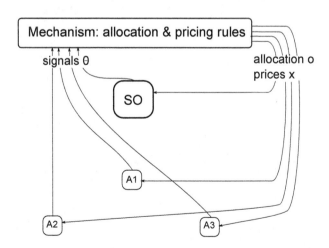

Fig. 1 Environment of a mechanism

include, among the other things, data for network modelling, market entities structure, commodities structure, offers, network system constraints, and individual constraints of the market participants. Figure 3 clarifies which data are inputs and outputs of the task.

In an elementary balancing task a market decision problem is solved. Usually it takes a form of an optimization problem. Assuming maximization of market economic surplus as a mechanism objectives, a general model solved in elementary balancing task is as follows:

$$\hat{Q} = \max_{p_o} [\sum_{o \in \mathcal{O}} K_0(p_o) - Z_0(p_o)] \tag{1}$$

$$\sum_{o \in \mathcal{O}} \alpha_{o,c} p_o = 0 \qquad \forall c \in \mathcal{C} \tag{2}$$

$$p \in P^0 \tag{3}$$

where

- $\mathcal{O}$ is a set of offers and o is an index offer, $o \in \mathcal{O}$,
- p_o is a volume of offer o accepted by mechanism,
- $K_0(p_o)$ is a cost of selling volumes p_o at offers prices,
- $Z_0(p_o)$ is a benefit of buying volumes p_o at offers prices,
- $\hat{Q}$ is a market economic benefit resulting from commodities exchanging,
- $\alpha_{o,c}$ is a commodity c ratio in an offer o,
- P^0 is a set of admissible allocations.

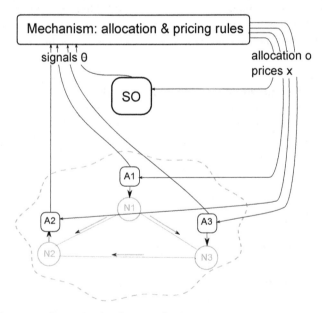

Fig. 2 Environment of a mechanism in networks system

A market surplus is an excess of the benefit of buying over the cost of selling volumes on the market. The objective (2) is a supply and demand balance for every commodity. Set P^0 arises from different system constraints, e.g. admissible offers volumes according to agents specifications or according to system constraints.

4 M^3 Architecture Overview

The architecture of the M^3 model is depicted in the Figure 4. The basis of M^3 lies in market mechanism layer. The layer describes the basic rules of market clearing apart from the structure of data and communication mechanisms which are specified in other layers.

Next layer, that is data model, describes all data needed to perform market processes in a real, productive environment or simulation and research environment. Here, general, conceptual model is specified by UML diagrams. M3-XML and relational model are some useful implementations of conceptual model. M3-XML model is destined mainly for signals in multi-agent environment and for ease to manage the cases – all required data can be packed into files and easily send to another system or research group. Model in relational database is suiteable e.g. for centralized data management of essential market processes.

Communication layer arranges the interactions between agents and solves some technical issues. Communication model can be described by UML sequence diagrams, e.g. comprising from communication acts based on some modifications of widely acknowledged FIPA (Foundation for Intelligent Physical Agents) standards. Possible communication models are certainly not limited to multi-agent approaches and FIPA standards, and can also be implemented e.g. in Web Services environment. The agents embedded in the market environment use the communication model to organize their interactions, to communicate with other agents and mechanism. During the communication acts they "talk" in M3-XML dialect using objects from data model layer. Finally, the mechanism can understand a language using M^3 data model and determines the allocations. Basic market processes are built on the three elements of M^3 framework: communication, data and mechanism.

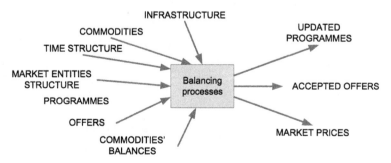

Fig. 3 Data in an elementary balancing task

5   M³ Formal Model

Balancing problem is defined by mechanism and also by market decision problem (MDP). A market decision problem describes whole situation providing data needed for market processes, including what commodities are, what agents are, what additional resources the agents needs and so on. A market decision problem can be solved with different balancing mechanisms, which leads to different balancing problem formulations. On the other hand, given balancing mechanism can be used for solving different market decision problems, which also leads to different balancing problem formulations.

Data related to a market decision problem are rather static. Dynamic values, such as offers, are passed by agents to mechanism and received from mechanism by agents as a signals. Finally, both the signals and market decision problem data constitute whole data being processed in market processes. M³ model embraces whole data in these two areas needed to solve balancing problem.

Space of market decision problems is characterized by a set of attributes. These attributes are defined in the area of system data and requirements (SDR) and market data and requirements (MDR). SDR should consists in data important from a non-market system point of view, like required additional resource. In this area, M³ defines data model for infrastructure reflecting commodities physical flow in the system. MDR defines data important from market system point of view, like time

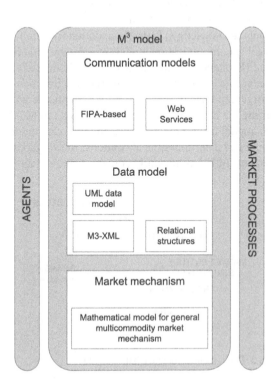

Fig. 4 M³ architecture

organization, commodities, agents, programmes (schedules). Finally, M^3 is used to represent offers.

5.1 System Data and Requirements Model

5.1.1 Infrastructure

Physical commodities exchange between market participants requires a technical infrastructure to assure feasible delivery of services and goods. Examples include telecommunication networks, roads and railways networks, transmission lines for the electricity, and so on. Thus the infrastructure plays the role of a system of limited resources during the balancing process. In the other words, infrastructure is a medium for delivery of commodities and services. However, a single balancing process may require different infrastructure models for different commodities. Moreover, the balancing process can integrate two aspects: pure trade operations and technical requirements to assure feasibility of trade operations. Both of them may require different infrastructure models, e.g. pure trade may be considered on national level, whereas feasible deliveries may need sub-regional, or zonal, resources. Therefore, many infrastructure models should be provided to balance the market. The most detailed infrastructure model is called a basic net.

Definition 1. (Basic network) Basic network is a tuple $n_0 = (V_{n_0}, E_{n_0}, (P_v^V), (P_e^E))$ where

- V_{n_0} is a non-empty set of nodes,
- E_{n_0} is set of ordered pair of nodes, $E_{n_0} \subseteq \{\{v_1, v_2\} : v_1, v_2 \in V_{n_0}, v_1 \neq v_2\}$,
- (P_v^V) is a vector of points from parameters space $\mathscr{P}^V$, $v \in V_{n_0}, P_v^V \in \mathscr{P}^V$,
- (P_e^E) is a vector of points from parameters space $\mathscr{P}^E$, $e \in E_{n_0}, P_e^E \in \mathscr{P}^E$.

Since other models of infrastructure aggregate the physical resources at some level of abstraction, we called them virtual nets.

Definition 2. (Virtual network) Virtual network is a tuple $n_0 = (V_n, E_n, (P_v^V), (P_e^E), \mathscr{V}_{\bar{n}})$ where

- $V_n, E_n, (P_v^V), (P_e^E)$ is defined as in definition 5.1.1,
- $\mathscr{V}_{\bar{n}}$ is a mapping between nodes of network n and nodes of related network $\bar{n}$, $\mathscr{V}_{\bar{n}} : V_{\bar{n}} \to V_n$; n_0 denotes a network being aggregated as a network n and for a given node $v \in V_{\bar{n}}$ mapping $\mathscr{V}_{\bar{n}}$ gives a node from V_n which aggregates node v.

The semantic of parameters (P_v^V), (P_e^E) depends on particular case and mechanism design. However, usually parameters related to edges represent the capacities and parameters related to nodes define admissible levels of commodities in a given node. Another typical usage is related to type of nodes and edges, e.g. sink nodes, source node. Definition of parameters spaces is an issue of mechanism designing.

We model infrastructure as set $\mathscr{N}$ of basic and virtual network models, where n_0 denotes basic network and $\mathscr{N} \setminus \{n_0\}$ is a set of virtual networks. There can be many aggregation schemas in parallel, however only relations between nodes in different nets are important, but not the relations between whole networks. Figure 5 shows the example of five networks and their aggregations.

5.2 Market Data and Requirements Model

5.2.1 Market Entities Structure

Market entities structure describes market agents and relation between them. Let us denote a set of all market entities (agents) as $\mathscr{A}$.

Definition 3. (Market entities structure) Market entities structure is a set of tuples $(a, \mathscr{R}_a, v)$ defined for every $a \in \mathscr{A}$, where

- $\mathscr{R}_a \subseteq \{(a, b) : a, a_1 \in \mathscr{A}, a \neq b\}$, ordered pair (a, b) denotes that entity a is in relation with entity b,
- v from network n, $\exists n : v \in V_n$ is a node in which entity a is located.

The market entities are modeled as an acyclic oriented graph, as shown in Figure 6.

Market entities form a hierarchy, where given market entity may be composed of some other market entities, e.g. a corporation and subsidiaries. Different market entities are involved at different stages of balancing process, e.g. on electrical energy markets each generation unit bids a sell offer, while the whole power plant is involved in settlements.

Also it may be convenient to introduce some artificial market entities. Such market entities may represent groups of players distinguished by certain feature,

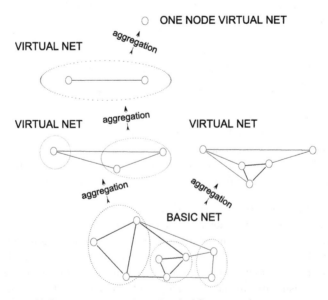

Fig. 5 Example of infrastructure model consisted of five networks

e.g. location in a given area, or production technology. Artificial market entities are useful to filter groups of market entities who require some additional resources.

5.2.2   Time Structure

Schedule of commodities' deliveries, which can be produced by the balancing process, is strictly related to time structure. Commodities are delivered in determined time slots. Thus the time horizon must be divided into time segments and every commodity is related to a time slot. However, trading different commodities may require different time slots, e.g. some commodity may be sold for an aggregated time slot, let's say - month, and some other commodities can be offered for shorter periods, say days, or hours. Moreover, a commodity can be offered to deliver only during the working days, peak hours, etc.

Figure 7 shows the example of time structure.

Definition 4. (Time structure) Time structure is a tuple $(\mathscr{H}, \mathscr{R}_{\mathscr{H}})$, where

- $\mathscr{H}$ is a set of all time slots required by mechanism (depending on balancing horizon and other design requirements),
- $\mathscr{R}_{\mathscr{H}} \subseteq \{(h_1, h_2) : h_1, h_2 \in \mathscr{H}, h_1 \neq h_2\}$ is a relation; if (h_1, h_2) are in relation, then time slot h_1 aggregates time slot h_2.

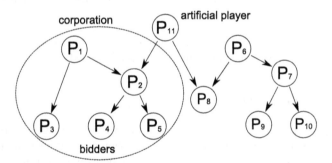

Fig. 6 Example of market entities

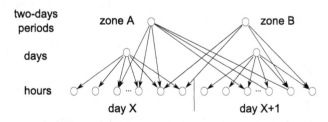

Fig. 7 Example of time structure

5.2.3 Commodities

Definition 5. (Commodity) Commodity c is a tuple $(v, h, P^{\mathscr{C}})$ or tuple $(e, h, P^{\mathscr{C}})$, where

- $v \in V_n$ is a node from some network n, in which commodity is defined,
- $e \in E_n$ is an edge from some network n, in which commodity is defined,
- $h \in \mathscr{H}$ is a time slot for which commodity is defined,
- $P^{\mathscr{C}} \in \mathscr{P}^{\mathscr{C}}$ is a point from parameters space $\mathscr{P}^{\mathscr{C}}$.

Let us denotes a set of all commodities as $\mathscr{C}$.

Observe that single commodity is always related to a single node or edge of the infrastructure model and a time slot. Nevertheless, it can be assigned to the root node of the most aggregated network, which can mean "the whole market". Thus the semantics is the following: "commodity is on the whole market and can be traded". The same way of thinking can be applied for time slots.

5.2.4 Programmes

Programme is a current schedule of delivery of some commodities for one or more market entities. Programme can be produced by balancing process as the result of accepting some offers. Programmes may be also an input data for some balancing processes. In this case it informs about the current overall status of the market entity contracts in an aggregated way.

Definition 6. (Programme) Programme is a function $S(a,c) : \mathscr{A} \times \mathscr{C} \to \mathbb{R}$, which for a given entity a and given commodity c returns contracted volumes.

5.3 Bidding Language

The agents present their preferences to the mechanism or other agents through offers. To evaluate the language they use, two issues should be considered. First issue is language expressiveness – what preferences can be expressed in the language. Second issue is the need to express the preferences in concise way. We propose offers model depicted in the Figure 8.

The first layer in the Figure 8 reflects the commodities. A simple offer is related to single commodity. Let $\mathscr{C}$ be a set of commodities.

Definition 7. (Simple offers) A simple offer is a tuple $o = (c, \alpha, s, d)$, where

- c is a reference to commodity c,
- α is a share factor $\neq 0$, positive for selling and negative for purchasing,
- s is an offer price,
- $d \subseteq \mathbb{R}$ is domain of offer volume, set of all admissible volumes.

When commodities should be offered in a given proportions, then bundle offers should be used. Bundled offers allows for submitting offers for packages (bundles) of different commodities simultaneously. Then a multicommodity mechanisms allow to allocate the resources according to some fixed proportions declared in the bundled offer. A participant submitting a bundled offer has to specify (i) proportions in which the particular commodities are supposed to be bought or sold in the resulting portfolio; and (ii) unit price of the set of bundled commodities in the portfolio. There is one offer price for the bundle. The market value of bundle is calculated as sum of market prices weighted by share factor. Since acceptance of bundled offer depends only on market value and offer price of bundle, there is not need for providing separate price for each product in the bundle. The M³ mechanism may accept the bundled offers partially, or entirely.

Definition 8. (Bundle offers) A bundle offer i is a tuple $o_i = (\{c, \alpha\}_i, s_i, d_i)$, where

- $\{(c, \alpha)\}_i, c \in \mathscr{C}1_i, \mathscr{C}1_i \subseteq \mathscr{C}$, where c is a reference to a commodity and α is share factor related to commodity c, $\mathscr{C}1_i$ is a subset of commodities appearing in offer i,
- s_i is an offer price,
- $d_i \subseteq \mathbb{R}$ is a domain of offer volume, set of all admissible volumes.

In real market trading some specific constraints or requirements may be required to be satisfied by sets of simple or bundle offers posed by participants. Grouping offers mechanism allows for incorporating such constraints into the decision model, so the decision can be performed under more realistic assumptions.

Grouping offers mechanism is a powerful tool in the multicommodity trading, that gives an extraordinary flexibility for market participants. The grouping offer submitted by an individual participant links a set of elementary and/or bundled offers, and provides specific constraints, such as resource requirements, between them. It allows to formulate individual constraints, or to bind elementary and/or bundled offers. Clearing with grouping offers can be realized in such a way that the total surplus is maximized under condition, that all grouping constraints are satisfied.

Let $\mathcal{O}_l^s$ be a set of simple offers and $\mathcal{O}_l^b$ be a set of bundle offers defined by agent l. Moreover, let $\mathcal{O}_l^{s,b}$ be a set of simple and bundle offers submitted by agent l.

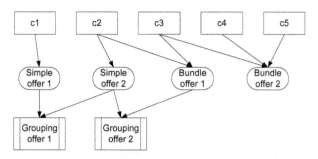

Fig. 8 Offers structure

Definition 9. (Grouping offers) A grouping offer defined by agent l is a tuple $g_i = (\mathcal{O}_l^{s,b}, \Phi_i, d_i)$, where

- $\mathcal{O}_l^{s,b}$ is a set of simple and bundle offers of agent l,
- Φ_i is an offer operator defined for volumes space of offers from set $\mathcal{O}_l^{s,b}$,
- d_i is a vector of admissible values for $\Phi_i(p_o)$.

Grouping offer i assures that allocation will satisfy the following constraints:

$$\Phi_i(p_o) \in d_i \qquad (4)$$

where p_o is a vector of accepted volumes for offers from $\mathcal{O}_l^{s,b}$.

6 Multi-commodity Balancing Model

Basic multi-commodity balancing model can be formulated as a linear programme as follows:

$$\hat{Q} = \max_{d,p} \left[Q = \sum_{m \in B} e_m d_m - \sum_{l \in S} s_l p_l \right] \qquad (5)$$

subject to

$$\sum_{m \in B} a_{im} d_m \leq \sum_{l \in S} \alpha_{il} p_l \qquad i = 1, \ldots, n \qquad (6)$$

$$0 \leq d_m \leq d_m^{\max} \qquad m \in B \qquad (7)$$

$$0 \leq p_l \leq p_l^{\max} \qquad l \in S \qquad (8)$$

where

- B is a set of buy offers,
- S is a set of sell offers,
- e_m is an offer price of buy offer m,
- s_l is an offer price of sell offer l,
- a_{im} is a share of commodity i in buy offer m,
- α_{il} is a share of commodity i in sell offer l,
- d_m^{max} is a maximal offered volume in buy offer m,
- p_l^{max} is a maximal offered volume in sell offer l,
- d_m is accepted volume of buy offer m,
- p_l is accepted volume of sell offer l.

In the objective 6 a market surplus is maximized. Left side of constraint (6) defines volume of realized demand, and right side defines volume of accepted supply. In this basic model it is assumed that balance is in a form of inequality, since it is easier to satisfy in a case of bundle trade. Thus, in general, accepted volume of sell offers can be higher then accepted volume of buy offers, if it is justified by market benefits maximization.

Distinction for sell and buy offer in the model is not unique and is not crucial for model. It is introduce to make the model more legible. In fact, in particular offer some commodities can be offered to sell while other to buy.

7 Summary

The chapter presents the overall design of the open Multi-commodity Market Model. We have formulated foundations of M^3 based on mechanism and game theory. We have formally defined elements of M^3. Introduced formal model can be transformed into data model to provide a language that agents can use for communication.

The proposed M^3 model has many potential practical applications. As it was shown, M^3 may be used in a wide range of market-oriented network systems and may significantly facilitate communication, coordination and modelling procedures, both from the market operators and market entities point of view. It may be used in designing information systems for market balancing and clearing in the context of multi-commodity trade in various network infrastructure sectors.

References

1. Hurwicz, L., Reiter, S.: Designing Economic Mechanisms. Cambridge University Press (2006)
2. Kacprzak, P., Kaleta, M., Pałka, P., Smolira, K., Toczyłowski, E., Traczyk, T.: M³ A common data model for diverse electrical energy turnover platforms. Rynek Energii 2(69), 12–18 (2007) (in Polish)
3. Kacprzak, P., Kaleta, M., Pałka, P., Smolira, K., Toczyłowski, E., Traczyk, T.: M³: Open multi-commodity market data model for network systems. In: Proceedings of 16th International Conference on Systems Science, vol. III, pp. 309–319 (2007)
4. Kacprzak, P., Kaleta, M., Pałka, P., Smolira, K., Toczyłowski, E., Traczyk, T.: Data Model Standarization for Real-Time E-commerce. In: Proceedings of the II International Multi-conference on Computer Science and Information Technology, pp. 579–588 (2007)
5. Toczyłowski, E.: Optimization of market processes under constraints, II extended edition. EXIT Academic Publishing (2003) (in Polish)
6. ebXML Web Site, http://www.ebxml.org
7. MDDL Web Site, http://www.mddl.org
8. RosettaNet organization website, http://www.rosettanet.org
9. XBRL Web Site, http://www.xbrl.org

M³ Data Structures

Tomasz Traczyk

Abstract. A detailed data model for multi-commodity trade is presented, based on formal model described in previous chapter. This data model has form of UML class diagrams, and widely uses generic (i.e. metadata driven) constructs. On the basis of the UML model, an XML dialect, called M3-XML, is proposed, which can be used to store formal descriptions of multi-commodity market and offers, or as a content language for market data interchange.

1 Introduction

As stated in previous chapter, existence of data models for open market information interchange systems is important for the future of network industries. M³ should therefore define a suitable data model and an appropriate communication language for multi-commodity markets [3].

A well-established good practice is to design the data structures in form of some implementation independent model. One of the most popular modeling conventions is UML with its class diagrams. This graphical modeling language ensures wide understanding of the model and has relatively large expressive power. Therefore, M³ data model [2] is primarily expressed as a set of UML class diagrams. Details of the model are described in Section 2.

One of the most important goals for M³ data structures is to establish a strong basis for creation of a flexible language, which is to be able to describe complex multi-commodity markets, commodities, and offers of market participants. Due to wide and well-justified popularity of XML, an XML dialect is proposed. The dialect, called M3-XML, is described in Section 3. This dialect should be suited for two important usages: as a content language for data interchange in electronic trade and as a complete formal notation of complex market problems for scientific research, optimization, etc.

Tomasz Traczyk
Warsaw University of Technology, Institute of Control and Computation Engineering
e-mail: T.Traczyk@ia.pw.edu.pl

M. Kaleta & T. Traczyk (Eds.): Modeling Multi-commodity Trade, AISC 121, pp. 21–46.
springerlink.com   © Springer-Verlag Berlin Heidelberg 2012

The data structures should be open and portable, so they should not depend on any proprietary solution or technology. Therefore, open standards are used for modeling and implementation of the structures.

2   M³ Data Model

A conceptual data model for multi-commodity market is proposed in form of UML class diagrams. The model describes the data necessary to express all concepts from the formal model (see chapter M^3 – *motivations and formal model*), supplemented with some additional data, useful for establishing communication, producing necessary trade documents, etc.

2.1   Assumptions

The data model for M^3 should be very flexible, because it should be capable enough to represent various complex models of diverse markets. Therefore generic solutions must be used, which enable adopting the data structure to various models by adjusting its metadata.

Though various markets can be modeled with M^3, a single "instance" of M^3 data structures represents only one market segment. Many instances of M^3 model may coexist for the same market segment, describing it from various points of view, e.g. created by different market participants. Thanks to common data structures and metadata, it should be possible, and relatively easy, to interchange information between such instances.

For clarity, an assumption has been made that the model concerns one balancing (decision making) cycle. Thus, the data determining validity periods of individual objects are omitted.

2.2   Data Structures

The data model is divided into several diagrams, representing the infrastructure (networks), market entities, commodities, offers and programmes.

2.2.1   Networks

A market may operate in network infrastructure. Figure 1 describes a data model for such networks.

A network consists of nodes and arcs. Both nodes and arcs can be of several kinds, which are characterized by different sets od parameteres. This is depicted in the diagram as exemplary sub-classes NetworkNode_ofKindX and NetworkArc_ofKindX. The structure is generic, i.e. metadata driven: possible kinds of nodes and arcs are defined by NetworkNodeKind and

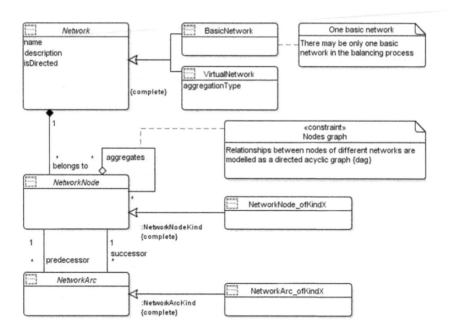

Fig. 1 UML class model for network structure

`NetworkArcKind` classes, respectively. This is shown in the diagram by means of UML powertypes[1].

Generic structure

The metadata for M³ generic structures is presented in Fig. 2 on page 24. A definition of each kind of object contains a name and description. Parameters for the defined object kind can be defined with their name, description, data type and unit of measure. A parameter may be required, which means that it must have a value for each instance of an appropriate object kind.

Node/arc kinds are defined by instances of metaclass `NetworkNodeKind`, and their parameters are defined by associated instances of `Network-NodeParameter` class. Classes called `*_ofKindX` are examples of classes defined by respective metaclass. Instances of these classes can have appropriate parameters, which values are stored in instances of `*Parameter` classes.

It may be necessary to use several models of the network for the same market segment:

- a basic network, which represents physical network or the least aggregated model of it (there may be only one basic network for each problem),
- virtual networks, which represent some more aggregated models of the network.

---

[1] A powertype (in UML 2.x [7]) is a metaclass, whose instances are subclasses of a given class; a powertype reference is denoted by a colon preceding the metaclass name.

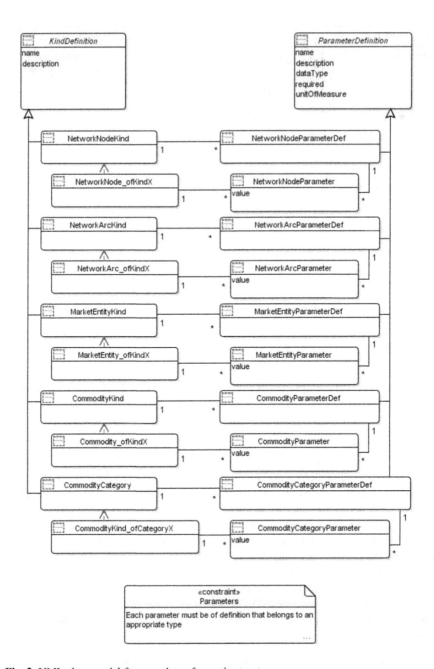

Fig. 2 UML class model for metadata of generic structures

Relationships can exist between network nodes on subsequent aggregation levels; these relationships are modeled as directed acyclic graph (abbrev. dag), represented by recursive association `aggregates` in the diagram.

2.2.2 Market Entities

Several subjects play various roles on the market – all of them we call market entities. The entities may be real business subjects (enterprizes, organizations), or may represent some virtual beings (created for the sake of balancing/decision making process), which can aggregate other entities. Appropriate data structure is shown in Fig. 3.

Market entities can be of several kinds, and can have various parameters; the kinds and parameters are defined generically, using metaclasses and powertypes, similarly to the solution described in Section 2.2.1; see Fig. 2.

An entity can be related to one or more network nodes. The relation can have various meanings (e.g. an entity can be physically located at the node, can be an owner of the node, can manage the node, etc.), this meaning is defined by the attribute `relationshipType` of the association class `EntityNodeRelationship`.

2.2.3 Commodities

There are many commodity kinds offered on the market. UML diagram that describes them is shown in Fig. 4. These kinds are divided into categories (e.g. storable

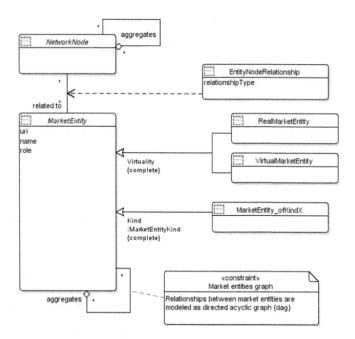

Fig. 3 UML class model for market entities

and non-storable commodities, options and derivative instruments, etc.). Both commodity kind and commodity category are defined generically (see Fig. 2). Thus, depending on its category, a commodity kind can be described by various sets of parameters.

`CommodityKind` class is a powertype for `Commodity` class, which means that each commodity belongs to one of subclasses of `Commodity` class (an exemplary subclass `Commodity_ofKindX` is shown in the diagram), but properties (name, etc., and parameters) of the subclass are defined by appropriate instance of the metaclass `CommodityKind`.

Real commodity is some good (or resource, or service) of given commodity kind, which is available at a specific period of time. The commodity may also be available at a specific node of the network (e.g. energy at power plant) or on a specific arc of the network (e.g. transmission rights). `activationTime` and `expirationTime` attributes define time period, when the commodity is available for market trading (this period may be, and usually is, different from period(s) of availability for delivery; see Section 2.2.4). The commodity can also be described

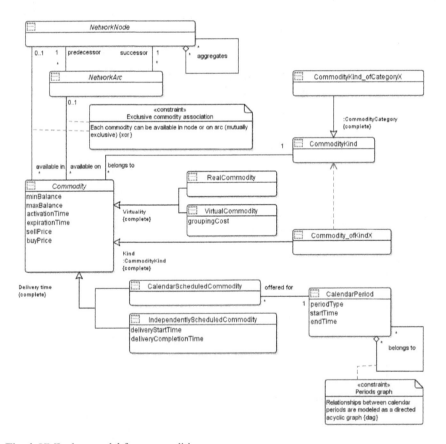

Fig. 4 UML class model for commodities

by a set of parameters, dependent of the commodity's kind. Virtual commodity is some abstract construct, used in grouping offers (see Definition 9 on page 18).

As a result of balancing/decision making process, a momentary balance must be obtained, implied as a difference between supply and demand, which must fit into constrains appropriate for given commodity (not necessarily zero); these constraints are defined by `minBalance` and `maxBalance` attributes. Sell and buy prices (which can differ – a difference may be used to cover costs of market operation) are also worked out in balancing/decision making process, and can be stored in appropriate attributes of `Commodity` class.

2.2.4 Calendar

M³ calendar defines set of time periods, in which commodities can be available for delivery. Such a construct is necessary on some markets, where various commodities (e.g. electrical energy at various network nodes) must be available in synchronized periods. A data structure for calendar is shown in Fig. 4.

Calendar periods can be aggregated into longer units; this is modeled as directed acyclic graph, represented in the diagram by recursive association `belongs to`. A commodity may be available (scheduled) in a set of calendar periods or in an independently defined period. `periodType` attribute classifies periods in a domain specific way (see examples on pages 59 and 61).

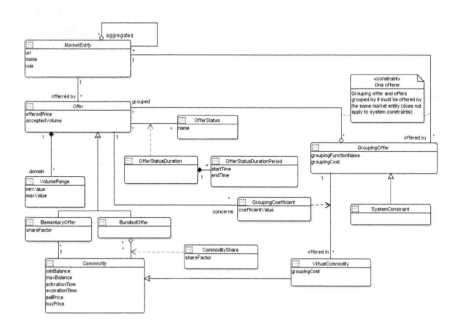

Fig. 5 UML class model for offers

2.2.5 Offers

A market is driven by buy/sell offers submitted by market entities and balanced in balancing/decision making processes. A data structure describing offers is presented in Fig. 5.

An offer contains offered price and defines an offered volume as a sum (union) of volume ranges. For sell offer, an offered price is a minimal price accepted by an offerer; for buy offer it is a maximal price, respectively.

An elementary offer contains only one commodity; a `shareFactor` attribute determines only whether it is sell (positive value) or buy offer (negative value).

Bundled offer contains many commodities, a `shareFactor` attribute determines fixed proportions between commodity volumes in the offered bundle (one offer may join selling and buying of commodities; it is determined by a sign of the `shareFactor`).

Grouping offer describes more complex combination of several elementary or bundled offers, bound by common conditions or shared resources. These offers are defined in terms of virtual commodities. Resultant acceptable volumes of individual commodities are calculated as some mappings of volume ranges acceptable for component offers. These mappings are defined by a matrix of coefficients, represented by `GroupingCoefficient` class. The same data model can be used to represent many system constraints, e.g. maximum power generated in one node of the network – see page 18.

An offer state can change, e.g. from 'published' to 'binding'. Possible states are defined by instances of `OfferStatus` class; periods (not necessarily disjunctive) when given status is current for given offer, are determined by association class `OfferStatusDuration` and associated class `OfferStatusDurationPeriod`.

When the offer is accepted in balancing/decision making process, an accepted volume of the bundle can be stored in `acceptedVolume` attribute of `Offer` class.

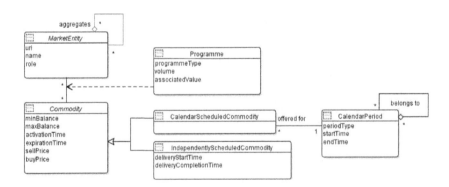

Fig. 6 UML class model for programmes

2.2.6   Programmes

Market states on input and output of balancing/decision making process in each balancing period can be represented by so called programmes (Fig. 6) – see Definition 6 on page 16.

Programmes store quantitative schedules of delivery/receipt of commodities by individual market entities.

3   M3-XML Language

As stated in previous chapter on page 5, an XML-based language representing M^3 data has been developed, which we call M3-XML. We propose two main applications of this dialect.

- Market data interchange between market participants. M3-XML language can be easily used as a content language in various communication architectures and protocols; see chapter *Communication models used in the context of multi-commodity trade*.
- Notation suitable for use in scientific experiments. The language enables researchers to precisely define complex trade models and scenarios. Since the XML-based notation can be easily transformed into other formats, such notation is well suited for cooperation with optimization solvers and other decision-making software. The notation is also self-explanatory and data files can easily be interpreted both by human and by programs. XML files can also be edited with use of standard text editors (like vi or Wordpad), so there is no need to create specialized software to maintain the data.

An information interchanged between market entities is to be expressed in XML, as a set of XML documents. Contents of these documents should reflect UML data model described in Section 2.

Though the communication models may differ, the main parts of messages should contain the same data, describing offers and/or balancing process results. So the XML documents' schemas for different communication models may contain the same body and differ only in envelope (headers and footers).

3.1   General Technical Solutions

3.1.1   Object Identification

In object-oriented modeling an assumption is usually made, that each object has its own identity. In object-oriented systems this identity is implicitly provided by the system; therefore in UML data model there is no need to bother with identifiers. Unfortunately, when we want to design XML-based data system, we must create explicit unique identifiers for objects.

A multi-commodity market is of course a kind of distributed system, which fact must be taken into consideration when we design unique identifiers.

- Unique identifiers must be system-wide, or better global; if they are not, a data interchange between system participants or between systems will encounter serious difficulties.
- If distributed system without any central entity is concerned (e.g. some kind of multi-agent system) there may be no possibility to create unique identifiers for some objects in a centralized way.

In M3-XML we propose a mixed strategy:

- Market-wide objects, like market entities, commodity kinds, calendar periods, must have unique identifiers created somehow in a centralized way. E.g. they may be created by some distinguished entity or may be imposed during system creation. The dictionaries defining such identifiers should be available for all market participants, e.g. they may be published as a Web service known to all participants. These dictionaries probably have to be replicated in participants' information systems, so this kind of unique identifiers creation is suitable only for objects that have relatively long lifetimes.
- Objects created by market participants should have unique identifiers created by their creators. To make these identifiers globally unique, they should be unique in namespace related to the creator, and – when used outside this namespace – should be qualified by unique identifier of the creator.

To fulfill this strategy, we decided to use URI (Unified Resource Identifiers) to build object identifiers, according to W3C standards [1], and to make use of XML Namespaces standard [8]. In M3-XML language, all attributes, which may contain object identifiers, are defined as of XML type QName (qualified name). It means that they consist of a local name qualified by URI of an appropriate namespace.

M3-XML-defined identifiers are created in M^3 namespace of URI `http://www.openM3.org/m3` (usually abbreviated to m3 by use of `xmlns:m3` attribute). Market/problem-wide identifiers should be created in a namespace of URI identifying the market or problem. Local identifiers, created by market participants, should be placed in their own namespaces, usually denoted by URL-s belonging to them.

3.1.2 Usage of XML Schema

M3-XML language is defined with use of XML Schema standard [9]. The definition of the language is divided into several documents, which share the namespace (as stated above), and import some common definitions stored in `M3GenericKinds.xsd` document.

In M3-XML schemas, a `<xsd:annotation>` elements are widely used to store detailed language components' descriptions, structured with specially designed sub-dialect. A language documentation presented on M^3 Web site [6] is generated automatically from these elements.

3.1.3   UML–XML Mapping

A mapping between UML model and M3-XML schema is quite straightforward, with the exception of generic structures (UML powertypes). Classes are represented by XML elements (with names starting with capital letters); attributes are represented as XML attributes (their names start with lowercase letters). Collections are represented by additional grouping elements (named by grouped elements' names in plural, in lowercase). Aggregation associations are represented straightly by XML hierarchical structure. Inheritance is represented by sub-elements corresponding to sub-classes. Another types of associations are represented by references (ref attributes) to elements' identifiers. Association classes are represented by additional intermediary elements (with names starting with lowercase letters), containing required attributes. Common components, such as repeated attribute sets or sub-elements, are represented as XML Schema reusable constructs, i.e. attribute groups and global complex types.

Generic structures, modeled as powertypes on UML class diagrams, must be represented in more sophisticated way. Each particular "Kind" metaclass is mapped into an appropriate XML element, which is used as a metadata defining a generic object kind. Concrete objects' data are stored in specialized elements, which are related to appropriate kind definitions by reference: dref attributes refer to "Kind" elements' id attributes. Details are described later, in Section 3.2.1.

3.2   M3-XML Schemas

Though schemas are XML documents, they are rather verbose and illegible, as they contain very detailed technical content. Therefore, it is reasonable to present some graphical representation of the schemas rather then schema documents themselves. So, in this chapter the schemas are presented in a graphical way, along with some basic description. More detailed description of the language and its current documentation can be found at our Web site [6].

3.2.1   Generic Definitions

Generic definitions are used widely in constructs of M3-XML, as it is designed to be very flexible language. Therefore the generic structures must be presented first.

Generic Types

A meta-metadata for M3-XML consists of XML Schema complex types, which are used to define metadata, i.e. generic types of objects. Definitions of generic object types (called 'kinds' to distinguish from XML Schema 'type'), which implement powertypes from the UML data model (Sec. 2, Fig. 2), are represented as elements of type KindDefinition_complexType (Fig. 9 on page 38). A typeParameter subelements refer (by their dref attributes) to definitions of object kind parameters – this construct is used for parameters, which are common for many object kinds. Parameters that are local to one kind, are defined locally by `ParameterDefinition` subelements. Common types and attribute groups are shown in Fig. 9, and are described later.

Definitions of instances of generic types (kinds) are represented as elements of type `GenericItem_complexType` (Fig. 8 on page 37). A `dref` attribute of the type refers to a definition of the object type (kind) – an element of `Kind-Definition_complexType` type, as described above. A `Parameter` subelements contain (in elements' text contents) values of instance parameters. These generic types are used to declare elements, which represent concrete kinds and instances of objects.

Definitions of commonly used auxiliary generic constructs are shown in Fig. 9 on page 38:

- generic groups are common attribute groups used in other constructs, e.g. for identification or description;
- `GenericAggregate_group` is used to represent object aggregation by a recursive reference (see e.g. Fig. 3) implemented by `ref` attribute;
- `dref` attribute is used to refer to an identifier (`id`) of an element, which specifies a metadata for appropriate type (kind) of object;
- `DomainDefinedQName_simpleType` is a qualified name (i.e. a name in its namespace) type restricted to refer to a term defined somewhere in domain dictionaries; a qualifier must be a URI of an appropriate dictionary (plain dictionary, taxonomy or ontology).

An example of M3-XML generic constructs can be found on page 188.

3.2.2 Networks

M3-XML `networks` element (Fig. 10 on page 39) is designed to define kinds of networks and their components (nodes, arcs) and to represent concrete instances of networks.

A `networkKinds` element contains specifications of node and arc kinds and their parameters. These specifications are created with use of generic meta-metadata types and common types described above, and they themselves constitutes a metadata for concrete instances of nodes and arcs.

Networks (basic and virtual, see 2.2.1) are described by `Network` elements (Fig. 11 on page 40). Nodes and arcs of a network are represented by `node` and `arc` elements; `dref` attribute of the elements refers to `id` attribute of an appropriate `NetworkNodeKind` or `NetworkArcKind` element. `Generic-Aggregate_group` is used to represent aggregation of nodes. `predecessor` and `successor` elements determine ends of network arcs. `VirtualNetwork` element can be used to denote a type of aggregation.

Examples of M3-XML network definition can be found on pages 184, 186 and 188.

3.2.3   Market Entities

marketEntities element (Fig. 11 on page 40) serves as a container for definitions of market entity kinds and data of concrete instances; the structure is similar to this described above.

A marketEntityKinds element contains definitions of the kinds and their parameters.

Concrete market entities are described with use of MarketEntity elements (Fig. 12 on page 40). dref attribute of the elements refers to id attribute of an appropriate MarketEntityKind element. GenericAggregate_group is used to represent aggregation of entities; uri elements determine URI-s of given market entity. VirtualMarketEntity element (if present) denotes virtuality of the entity.

Examples of M3-XML market entity definition can be found on pages 184 and 186.

3.2.4   Calendar

M3-XML calendar element contains set of CalendarPeriods, according to 2.2.4; appropriate XML structure is shown in Fig. 14 on page 42.

Examples of M3-XML calendar definition can be found on pages 59 and 61.

3.2.5   Commodities

commodities element (Fig. 16 on page 43) serves as a container for definitions of market entity kinds and data of concrete instances.

A commodityKinds element contains definitions of the kinds and their parameters, as usually.

Commodities are described with use of Commodity elements (Fig. 15 on page 15). dref attribute of the elements refers to id attribute of an appropriate CommodityKind element. Attributes of the element implement attributes of the commodity (see 2.2.3), and VirtualCommodity element (if present) denotes virtuality of the commodity. availableAt element's ref attribute refers to NetworkNode or NetworkArc element. CalendarScheduledCommodity or IndependentlyScheduledCommodity elements are used (mutually exclusive) to determine the commodity's period(s) of availability.

Examples of M3-XML commodities definition can be found on pages 59, 134, 134, 144, 185 and 186.

3.2.6   Offers

offers element (Fig. 17 on page 44) serves as a container for offers' data. It contains Offer elements, which describe elementary or bundled offers (see 2.2.5), and/or GroupingOffer elements describing grouping offers.

Offer description is quite complex, as shown in Fig. 18 on page 45. offeredBy subelement refers to MarketEntity element; offerStatus and `durationPeriod` subelements denote subsequent offer statuses and validity periods of the statuses. `volumeRange` subelements specify ranges of offered volume.

Elementary offer contain `ElementaryOffer` element of type `Offer-_complexType`, which refers to exactly one commodity (`ref` attribute of `offeredCommodity` subelement). In this case, a `shareFactor` attribute may only have +1 or -1 value, and determines whether this is a sell or buy offer. Bundled offer consists of many commodities, specified in subelements of `BudledOffer` element, and in this case `shareFactor` attribute determines proportion of the commodities in the bundle.

Grouping offer (Fig. 19 on page 45) consists of groups, containing commodity (`groupedCommodity` element) and grouped offer (`groupedOffer`). `coefficientValue`, `groupingFunction` and `groupingCost` attributes describe details of the grouping – see 2.2.5. `SystemConstraint` subelement (if present) determines that the grouping offer represents system constraint rather then real offer.

Examples of M3-XML offers can be found on pages 62, 63, 171, 172, 185, 187 and 189.

3.2.7 Programmes

M3-XML `programmes` element (Fig. 20 on page 46) contains a set of `Programme` elements, which describe programmes, according to 2.2.6. Each programme refers to a commodity (`ref` attribute of `programmeCommodity` subelement), and to a market entity (`ref` attribute of `programmeMarketEntity` subelement).

4   Applications of M3-XML Data Structures

As stated earlier, there are two main applications of M3-XML language: for market data interchange between market participants and as a notation suitable for use in scientific experiments. In this book we focus on the first application. In following chapters of the book we present possibilities to use M3-XML language in multiagent environment, in both centralized and distributed architectures (see chapter *Communication models used in the context of multi-commodity trade*). A research work on applications of M3-XML in ebXML-based centralized environment is also in progress [5].

M3-XML can be used in complex trade problems in many economy segments, including electricity market (see chapter M^3 *in the electricity market*), environmental protection (see chapter *Application of multi-commodity market model for greenhouse gases emission permits trading*), telecommunications (see chapter *Modelling virtual network market data with open Multi-commodity Market Model*) and airport capacity management.

5 Evolution of M^3 Data Structures and M3-XML Language

Since developed in 2007, M^3 data structures and M3-XML language have been be-ing successfully used by our and other cooperating with us research groups in many research activities, focused on (but not limited to) electricity market and telecom-munications. Though expressiveness and flexibility of the model and the language have been proven, some drawbacks have also been found.

One of known problems concerns verbosity of M3-XML representation of many real trade problems. Fortunately, in most cases many regularities can be noticed in such representation. In [4] we proposed two methods to shorten the notation of market data in M3-XML: a specialized extension of M3-XML language, and an approach based on generative programming and XVCL language.

Other problems concern lack of means of expression necessary for some very specialized cases of multi-commodity trade. These problems are being analyzed, and probably will be addressed in future versions of M^3 data model and M3-XML schema.

We expect important opportunities for further development of M^3-based market descriptions in use of ontologies along with M3-XML notation. In particular, we plan to define concepts and terms necessary to describe market segments, networks, commodities, market participants, etc., with use of ontologies and OWL language – see chapter *A SemanticWeb approach to the M^3 model*.

6 Conclusion

Designed as a flexible tool, based on well-thought-out mathematical model, and implemented as a modern XML dialect, a M^3 data structures seem to be convenient means to represent complex problems of trade, particularly in infrastructure sectors. M3-XML language has been successfully used in many our research activities, and – as we try to show in subsequent chapters of this book – can also become an adequate tool to use for market data interchange in difficult problems of electronic trade.

References

1. Berners-Lee, T., Fielding, R., Irvine, U.C., Masinter, L.: Uniform Resource Identifiers (URI): Generic Syntax. RFC 2396 (August 1998),
   http://www.ietf.org/rfc/rfc2396.txt
2. Kacprzak, P., Kaleta, M., Pałka, P., Smolira, K., Toczyłowski, E., Traczyk, T.: Data model for an open Multi-commodity Market Model M^3. In: Kozielski, S., et al. (eds.) Databases. New Technologies, pp. 289–300. WKŁ, Warszawa (2007) (in Polish)
3. Kacprzak, P., Kaleta, M., Pałka, P., Smolira, K., Toczyłowski, E., Traczyk, T.: Communi-cation Model for M^3 – Open Multi-commodity Market Data Model. In: Proc. 2nd National Scientific Conference on Data Processing Technologies KKNTPD 2007, pp. 139–150. Poznań (2007)

4. Kacprzak, P., Kaleta, M., Pałka, P., Smolira, K., Toczyłowski, E., Traczyk, T.: Notation methods for large volume regular data in complex electronic trade problems. In: Górski, A. (ed.) Information Systems Architecture and Technology, OWPW, Wrocław (2010)
5. Pałka, P., Traczyk, T., Wilk, R.: Centralized multilateral negotiations in multicommodity trade using the M^3 and ebXML standards. In: Proc. 7th National Conference Databases, Applications and Systems BDAS 2011 (in Press, 2011) (in Polish)
6. M^3 Web Site, http://www.openm3.org/
7. Object Management Group^{TM}: Unified Modeling Language^{TM}, http://www.omg.org/spec/UML/
8. World Wide Web Consortium: Namespaces in XML, http://www.w3.org/TR/xml-names/
9. World Wide Web Consortium: XML Schema, http://www.w3.org/XML/Schema

Appendix: M3-XML Schema

In this Appendix a graphical representation of M3-XML language schema is presented. Detailed schemas in XML Schema language can be found at our Web site [6].

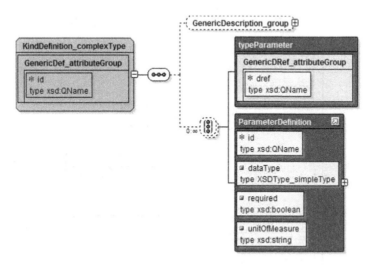

Fig. 7 XML Schema type definition for generic object kinds

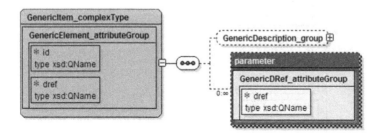

Fig. 8 XML Schema type definition for instances of generic object kinds

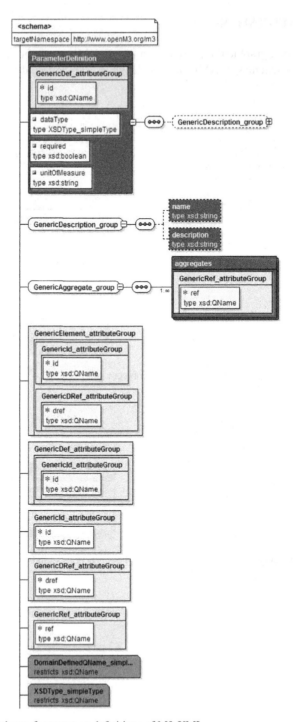

Fig. 9 XML Schema for common definitions of M3-XML

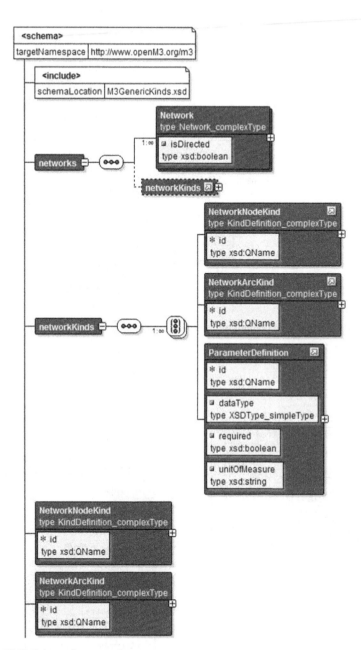

Fig. 10 XML Schema for networks

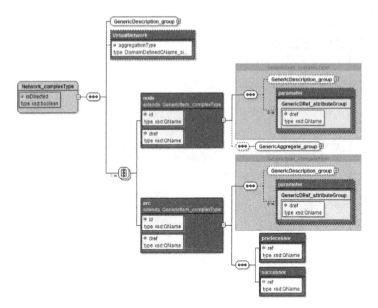

Fig. 11 XML Schema type definition for networks

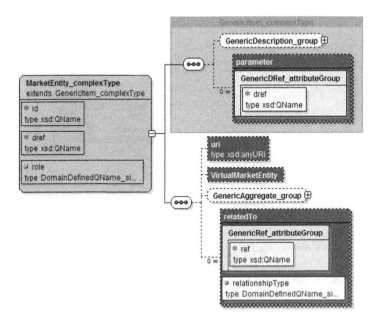

Fig. 12 XML Schema type definition for market entities

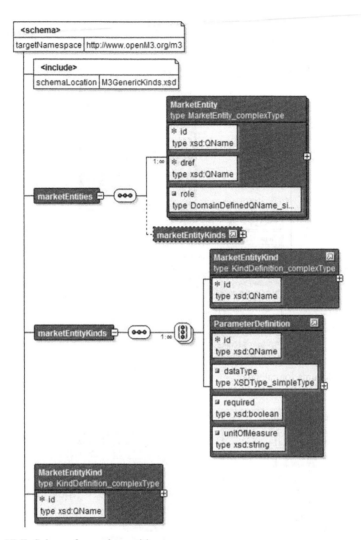

Fig. 13 XML Schema for market entities

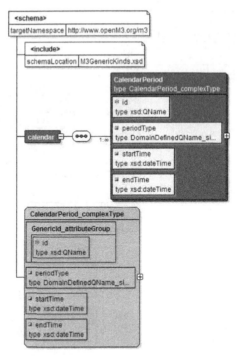

Fig. 14 XML Schema for calendar

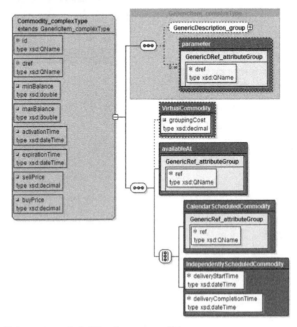

Fig. 15 XML Schema type definition for commodities

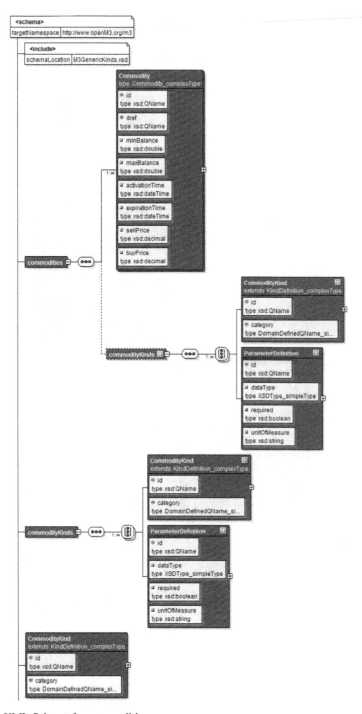

Fig. 16 XML Schema for commodities

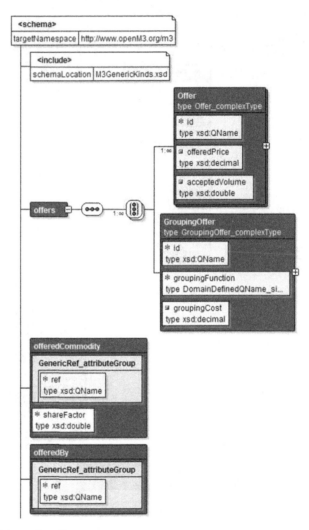

Fig. 17 XML Schema for offers

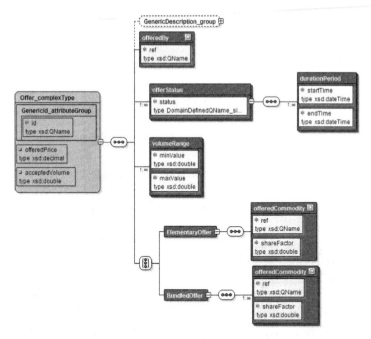

Fig. 18 XML Schema type definition for elementary and bundeled offers

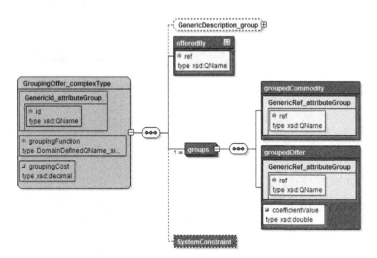

Fig. 19 XML Schema type definition for grouping offers

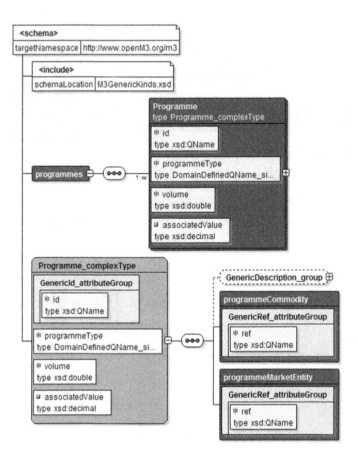

Fig. 20 XML Schema for programmes

Communication Models Used in the Context of Multi-commodity Trade

Piotr Pałka

Abstract. Multi-commodity trading can be performed both in centralized and distributed architecture environment. Choice of architecture creates a need to choose an appropriate communication model. Thus the definition of communication models is required. Foundation for Intelligent Physical Agents (abbrev. FIPA) organization, supplies us with communication models, appropriate for multi-agent systems. In this chapter we describe these models, and we also propose how FIPA standards can be enriched to multi-commodity and distributed elements.

1 Introduction

In this chapter we present trade models, associated communication architectures, and the agent communication language. Moreover, we present how FIPA standards can be enriched with multi-commodity and distributed elements. Finally, the details of agents communication language elements, used in multi-commodity trading, are presented.

2 Trade Models

To perform trading, first we have to make decision on the architecture in which the trade takes place. Most obvious architectures are distributed and centralized [3]. Both architectures have certain characteristics associated with the application of the communication model necessary to apply. Differences also lie in their potential applications.

Piotr Pałka

Warsaw University of Technology, Institute of Control and Computation Engineering

e-mail: P.Palka@ia.pw.edu.pl

M. Kaleta & T. Traczyk (Eds.): Modeling Multi-commodity Trade, AISC 121, pp. 47–64.

2.1 Distributed Markets

In the distributed market, participants (or agents) are directly engaged in exchange of goods, by negotiating the best, from their point of view, contracts. In such markets the various agents reach bilateral or multilateral agreements, usually after some, often complex negotiations. During negotiations, each negotiator may choose freely the negotiation partner, without the involvement of any central entity. Thus, in such markets, we observe the peer-to-peer communication (see Fig. 1).

The simplest type of agreement is a bilateral contract. Note, however, that in the case of multi-commodity trade (e.g. simultaneous energy and ancillary services trade), contracts may affect not only many goods, but also a number of suppliers and customers. Such agreements is called multilateral.

Distributed market architecture could be represented as a network communication architecture. Many negotiators, as well as many independent brokers are associated with the distributed market architecture. Negotiators communicate with each other in peer-to-peer manner. Brokers collect information from particular negotiators and try to balance considered offers.

2.2 Exchange-Like Markets

In this class of the markets there exists a central agent (e.g. the market operator), which collect offers sent by other agents involved in the trade. Such agent balance the demand and supply on the market, according to some criteria. The central agent distributes the balance results to all agents involved (see Fig. 2). It may also ensure compliance with various restrictions, e.g. physical constraints associated with the production and transmission of electricity.

The simplest type of such mechanism is the stock market, where there are no physical limitations for delivery. The bids can be ordered by price and most favorable offers are accepted, while others are rejected. In the simplest case, the equilibrium price and total volume are determined in the intersection of supply and demand curves. Such rules are used e.g. in the wholesale electric power exchanges. However, there is often a need to reflect the above-mentioned physical limitations in the balancing processes.

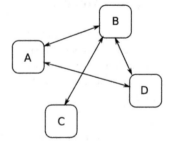

Fig. 1 Distributed market concept: A, B, C and D are market participants, communicating on a peer-to-peer manner

With the centralized trading architecture, the star-like communication architecture is related. In centralized market architecture usually one central agent is associated. It performs centralized balancing processes and controls data flow between traders. Synchronous balancing is mostly assumed with such communication architecture.

2.3 Morris Column Agent

In the centralized system it is possible to create a central repository, which collects all the necessary data about the agents, and existing and planned trade processes. As distributed systems do not have (and often cannot have) any central entity, such repository cannot be built, so there is a need to share the information in a different way. For this purpose we define a special agent called Morris Column agent [4]. The task of such agent is to offer some public location, where other agents may "hang a notice" to report certain information about trade processes provided by them. In addition, the Morris Column agent should provide a functionality of searching and removing the information. In a broader context, agents may leave notices on initiating various types of auction processes with some market-specific characteristics.

In the chapter *Application of the multi-agent systems in the context of M^3* we propose the Morris column agent implementation.

2.4 Physical vs. Economical Aspects

With complex market architecture many different market processes are associated. There is a need to develop proper choreography of different processes in whole market. Moreover, different market processes, which relate to the same commodity, operate for various distances from the time of commodity delivery. Example of such a market could be the electricity energy balancing market, with the day ahead market, hour ahead market and the real time market. Each of the market processes is characterized by a different distance in time from the energy delivery, as well as different set of constraints. These constraints, with the approaching date of delivery, are becoming more detailed. Different aspects of trading play a crucial role on different market processes, on the one hand physical aspects (e.g. need to ensure

Fig. 2 Centralized market concept: A, B, C and D are market participants, involved in trading. SO is a central agent, collecting offers, balancing supply and demand according to market mechanism

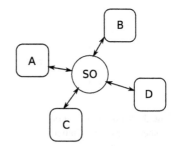

infrastructure constraints on real time market), on the other hand economic aspects (e.g. more emphasis on price on the day-ahead market). These relationships are shown in Fig. 3.

Generally, when the commodity delivery date is imminent, the physical aspects are essential. The physical constraints bind different market participants. There is a need for supervisor (often system operator) which cares about the technical constraints satisfaction. These requirements make it necessary to apply centralized trade.

On the other hand, when a lot of time remains to date of commodity delivery, the economic aspects are more important. Often the physical constraints are not considered, or are less important. On such markets most often the bilateral (or multilateral) contracts are concluded. Thus the distributed trade is possible.

The market segment localization on the timeline determines both the importance of physical or economic aspects, and the target market and the communication architecture. The distributed architecture is more suitable, when it is enough time to delivery time slot – economical aspects are more important. The centralized architecture is more suitable, if it is closer to delivery time slot – physical aspects are more important.

3   Evolution of Agents Communication Languages

Agents communication language (abbrev. ACL) defines how agents can exchange information and can conduct trade processes. ACL determines how agents interact, and thus determines also how agents exchange messages with each other. During agents' interaction some content is passed.

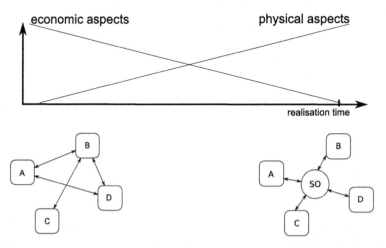

Fig. 3 Physical vs economic aspects, depending on market segment localization, importance of specific aspects changes

3.1 Speech Act Theory

Speech act theory aims at a comprehensive analysis of speech acts and their typology. Speech act is a message sent by the sender to the recipient, which is intended to convey a message by using a system of language signs.

The speech act theory was initiated by the work of John Austin [1]. In this book he distinguished three different aspects of speech acts. As the performance of the utterance (the *locutionary act*); the action performed in making the utterance – the "illocutionary force" of the utterance (the *illocutionary act*); and in certain cases an actual effect of the utterance (the *perlocutionary act*) [9].

Consider that sender sent a request to receiver to perform an action. In [8] author identifies several properties that must hold for a speech act performed between a sender and a receiver to succeed. The normal I/O conditions, so that the receiver has to receive the request, have to be fulfilled. Also the preparatory conditions must be met, so the sender has to choose the act correctly, receiver must be able to perform action, and the sender must believe that receiver is able to perform action. Moreover, it must not be obvious, that receiver will do action anyway. Finally the sincerity conditions, which state, that the sender really wants an action to be performed [9], must be fulfilled.

In [8] the systematic classification of speech act is performed:

- representatives (informing), which commit the sender to the truth of an expressed message;
- directives (requesting), in which the sender requests the receiver to perform requested action;
- commisives (promising), which commit the sender to a course of action;
- expressives (thanking), which express some psychological state;
- declaratives (declaring), which effect some changes in a state of affairs.

Performatives

Most important for us is to identify which of the speech acts may be considered only in terms of effectiveness, not the veracity. These speech acts are called performatives. The entities (agents, market participants), communicating with each other, use performatives just because they can only assess the message effectiveness, not its accuracy.

3.2 KQML

One of the first agents communication languages, the Knowledge Query and Manipulation Language (abbrev. KQML), was developed in 1990s [2]. The KQML is based on message passing. It defines "envelope" of the message, which contains important information about message (e.g. sender and receiver identifiers, used language, ontology etc). Each message is described by the performative, which

defines the class of message – the communicative act. The performative defines
communicative act desired interpretation. Exemplary communicative act is presented on Listing 1.

Listing 1 Exemplary KQML communicative act

```
(ask-one
 :content (PRICE ENERGY ?price)
 :receiver energy-broker
 :language LPROLOG
 :ontology energy-market
)
```

The KQML communicative acts set is very fuzzy. Different implementations
of multi-agent systems often has problems with cooperation because of undeter-
mined meaning of particular performatives, and also because of different interpre-
tations of the same performative. Moreover, the mechanism of message passing
between two agents has never been precisely defined, so there are difficulties in
agent communication.

Also, the KQML semantics has never been precisely defined, thus agents never
use "proper" KQML communication language – the messages could have different
interpretation. Moreover, KQML has never defined whole class of commisives acts,
so it is impossible to send, e.g. a commit message. Finally, the KQML communicative acts set is very large – according to [9] it contains about 40 communicative
acts.

Because of many drawbacks listed above, KQML communicative language is
a bad solution. Therefore, let us look closer at the communicative language, proposed by the FIPA organization in 1999. This language is based on KQML, nevertheless takes account of defects listed above and corrects them.

3.3   FIPA

Foundation of Intelligent Physical Agents (abbrev. FIPA), is an organization that
promotes use of multi-agent environments. FIPA has developed several standards
for multi-agent communication, has clarified Communicative Acts, an has identified and proposed a number of interaction protocols. The main frame is the Agent
Management Reference model. In this model, every agent is embedded in a standardized platform, which provides the main system for agents communication. On
this platform, the agent must operate with awarded AMS (Agent Management Service) that manages a registry of all agents acting within the platform. In addition,
the DF (directory facilitator) agent can be defined, which informs agents about existence of other agents and services they provide [10].

Agents communicative language FIPA-ACL is syntactically very similar to
KQML language. The envelope structure is syntactically the same, attributes are
also very similar. It is obvious, since both languages are based on the speech act
theory. In designing the FIPA communicative language, KQML deficiencies have

been taken into account. FIPA-ACL defines 20 performatives, each of them has precisely defined meaning.

FIPA standards can be enriched with multi-commodity and distributed elements in accordance with the M^3 model [5, 7]. The analysis performed in [7] shows that it is worth to extend some elements of the FIPA standards, in particular, interaction protocols and the content language of the messages, in order to obtain better conditions for the exchange (such as the possibility of multilateral negotiations, or trade in the distributed environments), and get fit for the M^3 model. By combining the M^3 model, which offers a rich modeling capabilities for multi-commodity restricted markets, with the widely used communications standard, we obtain wider opportunities for modeling markets.

4 Agents Communication Language

Agents Communication Language (abbrev. ACL) consists of Communicative Acts, Interaction Protocols, and Content Languages. A method of interpreting the content of the message is defined by three elements: the content language, the performative and the ontology used for exchange of information. The content language specifies a syntax of message content, whereas performative contained in the communicative act defines a semantics, and thus gives meaning to the constituent content of messages. The ontology completes the semantics.

The Interaction Protocols are defined as the collection of relevant Communicative Acts, which are exchanged sequentially between particular agents. Each Communicative Act has a content, which meaning is defined by the Content Language specification (see Fig. 4).

4.1 Communicative Acts

Single communicative act contains some portion of control data, e.g. sender's and receiver's identifiers, type of the message (including semantic meaning), etc. This is defined in FIPA specifications, so we do not consider it in this chapter.

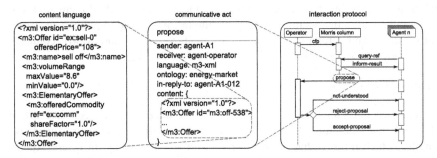

Fig. 4 Dependencies between interaction protocols, communicative acts and content languages

Each communicative act has a certain performative associated. Such performative defines how the content should be interpreted (semantically).

4.2 Interaction Protocols

During the multi-agent communication we can observe some typical patterns of message flow. FIPA has collected these patterns and called them Interaction Protocols. From the standpoint of a multi-agent trading platform designer, an interesting protocols are *Contract Net Interaction Protocol* and *Iterated Contract Net Interaction Protocol*. They define the interchange of messages in an auction-type exchange or in negotiation based markets.

Contract Net Interaction Protocol defines messages interchange for centralized trade (in auction or exchange). Agent, which submits its offer using this protocol, has no possibility to negotiate. Iterated Contract Net Interaction Protocol is an extension of the preceding one. This protocol assumes that offers may be negotiated, so it can be used on markets, where negotiations are necessary, e.g. for bilateral or multilateral trading. In the following sections our extension of these two protocols is presented, which enables them to be used for multi-commodity trade.

4.2.1 Centralized Markets

On the centralized market the system operator exists, that is responsible, among other things, for creating and maintaining a platform, where market participants can submit their sell and buy offers. The operator specifies the market model (under some regulation rules), determines the range of traded commodities, their characteristics, the rules for transaction making and clearing, infrastructural (e.g. network) constraints, other system requirements and constraints, etc. The operator usually defines all necessary dictionaries (sometimes defined in the form of ontologies): lists of commodity kinds, market entities, etc. The operator defines also a calendar, which determines time periods in which commodities can be delivered. Agents willing to trade on the centralized market must know where such platform can be found and how it can be used. If such a platform is not commonly known, it must be somehow announced to interested agents. It is of course impossible to notify all potentially interested agents directly, because they are numerous, and much of them are probably not known to the operator. A Morris column can be used to publish necessary information. An operator submits announcements to one or more Morris columns, where they become available to all interested agents (see Fig. 5).

Modified Contract Net Interaction Protocol

A sequence diagram in Figure 6 presents a modified Contract Net Interaction Protocol. The operator submits a cfp (*call for proposal*) message to Morris column(s), which informs that the platform is available and waits for offers. Agents scans commonly known Morris columns, looking for interesting trade opportunities. They use *query-ref* messages that contain patterns specifying, which information is interesting to the sender. A Morris column responds to these queries sending *inform-result* messages, which contain only information fitting the patterns.

Agents interested in trade on the particular platform submit their offers, using *propose* message, to the platform announced on the Morris column. Depending on market model, offers should be submitted in specific time window, or can be submitted at any time. If the offer is not correct (e.g. it can not be understood or is incomplete), a *not-understood* message is sent to the offeror. The operator joins the offers using algorithms specific to the market model. Next, *accept-proposal* messages are send to the offerors, whose offers succeeded, and *reject-proposal* are send to the agents, whose offers failed. In this kind of market an offer is binding – an agent, which submits the offer, is obliged to fulfil the commitment.

4.2.2 Decentralized Markets and Negotiations

Decentralized markets lack of a central institution, where offers can be placed, and which determines the rules and common dictionaries for the market. Such markets

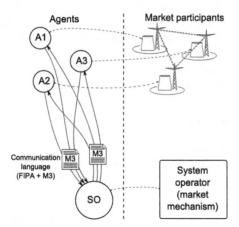

Fig. 5 Schema of agents' interaction on the centralized market and the dependencies between agents and market participants; agents are using the proposed communication language

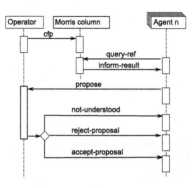

Fig. 6 Modified Contract Net Interaction Protocol

are based on bilateral or multilateral trade, but before any trade can take place, potential participants must find each other. One of possible methods is the use of Morris columns, on which agents place their announcements.

Modified Iterated Contract Net Interaction Protocol

Figure 7 shows proposed scenario of such bilateral interaction. Agent submits a *call-for-proposal* message to one or more Morris columns. The message contains detailed description of the offer: commodities, quantities, time windows, etc. Other agents look for interesting offers, sending *query-ref* messages to appropriate Morris columns. As a result of the query they obtain an *inform-result* messages containing lists of offers that fit the query; these lists contain also URL addresses of agents-offerers. Next, agents enter into direct bilateral or multilateral negotiations. They send *propose* messages to each other in one ore more iterations, negotiating the terms of trade. If negotiations succeed, one of the agents sends *accept-proposal* message the other one.

Unfortunately, this quite simple schema does not work if any of the parties continues searching for better offers simultaneously to negotiations. Offers on this type of market usually are not binding, and each party can break off the negotiations at any time, whenever it finds a more profitable solution, even when the negotiations are almost finished. So, submitting *accept-proposal* message does not confirm the transaction, it must be additionally committed using special protocol, which ensures that all the parties involved in the transaction accept the result of the negotiations.

4.2.3   Two-Phase Commit

An adaptation of two-phase commit protocol (2PC), well known in distributed OLTP (On-line Transaction Processing) systems, can be used as a method for

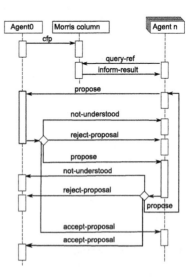

Fig. 7 Modified Iterated Contract Net Interaction Protocol

accepting transactions in decentralized markets. It guaranties safe completion of transactions, ensuring that all involved parties ultimately accept the contract. Figure 8 presents a two phase commit protocol.

Unfortunately, FIPA standards do not directly define this protocol, but it can be quite easily defined on base of communication acts existing in the standards. In [6] an appropriate communication schema is proposed for bilateral transactions, but the 2PC protocol can also be used for multilateral ones.

Agent, which initiates completion of the transaction, sends *request(global-commit)* message to the other participants of the transaction. The parties should answer using *accept(global-commit)* message if they accept the transaction, or *reject(global-commit)* otherwise. If all of the parties accept the transaction, an initiating agent sends *confirm(global-commit)* message and the transaction is irrevocably accepted. If any of the parties rejects the transaction, a *disconfirm(global-commit)* is send and the transaction fails. If any of the parties does not answer in a given time, the transaction is timed-out and it also fails.

4.3   Content Languages

The main content of the message is interpreted differently, depending on used content language, performative and ontology [10]. FIPA proposes a set of formal semantic content languages. The most refined is the FIPA-SL (FIPA Semantic Language). However, semantic language can be awkward tool to write market data, especially if we want to send a large, internally related collection of data. According to the FIPA standard, the content language should be comprehensible to each of agent participating in the communication. This is justified by the fact that the agents should understand each other. Thus, assuming that the agents have implemented methods to read data stored in given language, we propose the new content language, based on the M3-XML format.

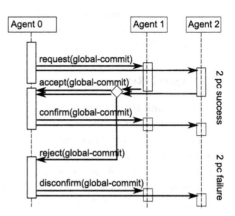

Fig. 8 Two Phase Commit Protocol

4.3.1 Call for Proposal (abbrev. cfp)

This communicative act is used to report willingness to negotiate, or to the declaration that some kind of market/auction begins. The *cfp* communicative act should include following elements.

- A unique identifier of the market/negotiations.
- Type of trade/negotiations. This may be a market contract (on which agents will negotiate), or the stock market/auction. For the second possibility we consider the continuous or uniform trading.
- Time window in which agents should submit bids/proposals A single time window is defined by a `m3:calendarPeriod` element. A collection of time windows is surrounded by a `m3:calendar` element.
- List of commodities which the agent wants to trade. A single commodity is defined by a `m3:Commodity` element. A collection of commodities is surrounded by a `m3:commodities` element. The definition of single commodity includes:
  - period(s) in which the commodity is available;
  - in the case of commodity associated with certain infrastructure network, an association with a network resource (node or arc);
  - a number of additional parameters, whose meaning is defined by an ontology associated with that type of trade.

We propose the following notation of these elements, using the XML notation (due to the fact that some items, as the commodities or the time windows, are defined by means of XML-based language M3-XML). An exemplary structure of the message is presented on Listing 2.

Attribute `Type` determines whether we are dealing with auction (`type="stock"`) or negotiation (`type="Negotiations"`). When we deal with auction, an attribute `Quotation` determines whether there is a single-price auction system (`Quotation="single"`) or continuous (`Quotation="continuous"`). In the event that we are dealing with negotiations, we need to determine whether the negotiations will focus on sales (`OfferType="ask"`) or purchase (`OfferType="bid"`).

4.3.2 Query-Ref

Such communicative act is used by agents to submit a query to Morris column. Agent, which wants to obtain a list of advertisements, sends a query to the Morris column. Since the advertisements are XML documents, the most appropriate tool is XPath language [11] or XQuery language [12].

Let us assume that agent wants to query the Morris column about all advertisements, which comprise selling electric energy in negotiations process. This query notated in XPath language is presented on Listing 3.[1] In the XQuery

---

[1] For simplicity, declarations of appropriate namespaces (e.g. m3 namespace) are omitted.

language, this query (assuming that the document that stores the announces is called
`MorrisColumnAnnounces.xml`) is presented on Listing 4. Thus, if the agent
reaches the XPath language, the *query-ref* message is presented on Listing 5, or,
if the agent formulates a query in XQuery language, the message is presented on
Listing 6.

Listing 2 Exemplary structure of the *cfp* message

```
(cfp
 :sender ex:Agent0
 :receiver ex:MorrisColumn01
 :content (
      <market id="ex:001" type="stock" quotation="continuous">
        <m3:calendar>
          <m3:CalendarPeriod id="ex:Y2009"
                             periodType="ex:one-year"
                             startTime="2009-01-01T00:00:00"
                             endTime="2010-01-01T00:00:00"/>
        </m3:calendar>
        <m3:commodities>
          <m3:Commodity dref="op:ElectricEnergy"
                        id="ex:eY2009">
            <m3:description>
              Energy on Copper Plate
            </m3:description>
            <m3:availableAt ref="op:CPNode" />
            <m3:CalendarScheduledCommodity ref="ex:Y2009" />
          </m3:Commodity>
        </m3:commodities>
      </market> )
 :language m3xml
 :ontology EnergyMarket )
```

Listing 3 Exemplary query notated in XPath language

```
/announces[/market[@type='negotiations' and @offerType='ask'] \
/m3:commodities/m3:Commodity[@dref='op:ElectricEnergy']]
```

Listing 4 Exemplary query notated in XQuery language

```
for $x in doc("MorrisColumnAnnounces.xml")
  where $x/announces/market/@type='negotiations'
    and $x/announces/market/@offerType='ask'
    and $x/announces/market/m3:commodities/ \
        m3:Commodity/@dref='op:ElectricEnergy'
return $x
```

4.3.3 Inform-Result

The content of this communicative act is a list of the announcements. As the advertisements are formed in the XML language, the answer contains a list of inform-
results; each result is related to particular XML advertisement. Stucture of the message is presented on Listing 7.

Listing 5 Exemplary structure of the *query-ref* message

```
(query-ref
 :sender ex:Agent2
 :receiver ex:MorrisColumn01
 :content ([/announces/announce/market[@type='negotiations' \
           and @offerType='ask']/m3:commodities/ \
           m3:Commodity[@dref='op:ElectricEnergy']] )
 :language xpath
 :ontology EnergyMarket
 :reply-with xpath:query02 )
```

Listing 6 Exemplary structure of the *query-ref* message

```
(query-ref
 :sender ex:Agent3
 :receiver ex:MorrisColumn01
 :content (
    for $x in doc("MorrisColumnAnnounces.xml")
      where $x/announces/announce/market/@type='negotiations'
        and $x/announces/announce/market/@offerType='ask'
        and $x/announces/announce/market/m3:commodities/ \
            m3:Commodity/@dref='op:ElectricEnergy'
    return $x
 )
 :language xquery
 :ontology EnergyMarket
 :reply-with xpath:query03 )
)
```

4.3.4 Propose

This act contains an ask/bid (proposal in the negotiation process may also be regarded as a type of a deal). Thus, for this communicative act we propose to use M3-XML as a language for defining offers. An example of a message is presented on Listing 8.

Communication act is interpreted as a proposal sent by the agent ex:Buyer to the agent ex:Operator in m3 language (i.e. in the M3-XML language). The content of the act has the following meaning: the offer written in M3-XML language is a buying offer (ShareFactor="-1") of the commodity ex:energyOn2007-04-09T08:00:00 (detailed description of this commodity may be described by the ontology EnergyMarket, or by an additional element of M^3, which is the description of commodities). Agent ex:Buyer sends the unit price offeredPrice="-140.00" (minus sign is conventional in the M^3 model) and the quantity of this commodity, ranging from minValue="0" to maxValue="100". Offer has a unique identifier id="ex:o12345-67".

4.3.5 Reject-Proposal

This act is designed to notify the parties of the rejection of bid/proposal previously submitted by that party. Bid/proposal id is currently defined by the element Reply-with.

Listing 7 Exemplary structure of the *inform-result* message

```
(inform-result
 :sender ex:MorrisColumn01
 :receiver ex:Agent2
 :content (
   <announces>
     <announce agentId="ex:Agent8">
       <market id="ex:004" type="negotiations" offerType="ask">
         <m3:calendar>
           <m3:CalendarPeriod id="ex:W23-29032009"
                              periodType="ex:one-week"
                              startTime="2009-03-23T00:00:00"
                              endTime="2009-03-30T00:00:00"/>
         </m3:calendar>
         <m3:commodities>
           <m3:Commodity dref="op:ElectricEnergy"
                         id="ex:eM062009">
             <m3:description>
               Energy on Copper Plate
             </m3:description>
             <m3:availableAt ref="op:CPNode"/>
             <m3:CalendarScheduledCommodity
                         ref="ex:M062009"/>
           </m3:Commodity>
         </m3:commodities>
       </market>
     </announce>
     <announce agentId="ex:Agent32">
       <market id="ex:007" type="negotiations"
                           offerType="ask">
         <m3:calendar>
           <m3:CalendarPeriod id="ex:D01042009"
                      periodType="ex:one-day"
                      startTime="2009-04-01T00:00:00"
                      endTime="2009-04-02T00:00:00"/>
         </m3:calendar>
         <m3:commodities>
           <m3:Commodity dref="op:ElectricEnergy"
                         id="ex:eD15052009">
             <m3:description>
               Energy on Copper Plate
             </m3:description>
             <m3:availableAt ref="op:CPNode"/>
             <m3:CalendarScheduledCommodity
                     ref="ex:D15052009"/>
           </m3:Commodity>
         </m3:commodities>
       </market>
     </announce>
   </announces>
 )
 :language m3xml
 :ontology EnergyMarket
)
```

There are the following reasons for rejection of bid/proposal:

- out-of-time – the offer was received outside the period of trade/negotiation;
- incorrect-commodity – the offer was made to the wrong commodity;
- price-unacceptable – the proposed price was unacceptable;
- negotiations-expired – negotiations were expired.

Listing 8 Exemplary structure of the *propose* message

```
(propose
 :sender ex:Buyer
 :receiver ex:Operator
 :content (
      <m3:Offer id="ex:o12345-67" offeredPrice="-140.00">
       <m3:description>Buy offer</m3:description>
       <m3:offeredBy ref="ex:Buyer"/>
       <m3:volumeRange minValue="0" maxValue="100"/>
       <m3:ElementaryOffer>
        <m3:offeredCommodity shareFactor="-1"
                      ref="ex:energyOn2007-04-09T08:00:00"/>
       </m3:ElementaryOffer>
      </m3:Offer>)
 :language m3xml
 :ontology EnergyMarket
 :reply-with ex:o12345-67)
```

Listing 9 Exemplary structure of a *reject-proposal* message

```
(reject-proposal
 :sender ex:Operator
 :receiver ex:Buyer
 :content (negotiations-expired)
 :language sl
 :ontology EnergyMarket
 :in-reply-to  ex:o12345-67
)
```

A sample message which reject the offer because of the lack of a product is presented on Listing 9 – the agent has completed negotiations with another agent and sold him the commodity.

4.3.6 Accept-Proposal

The *accept-proposal* act is designed to inform the agent who previously submitted a bid/proposal that it was accepted. The message contains an output offer (it could be accepted partially). Exemplary structure of a *accept-proposal* message is presented on Listing 10.

4.4 Ontologies

The parameters of trade conditions are mainly passed as M3-XML documents. These relate to a range of commodities, entities, infrastructure networks, and calendar structure. Some of this data are expressed by references to open dictionaries, defining kinds of commodities, kinds of entities, etc. The dictionaries must be defined for each trade process independently. If it is possible the definition of a broad dictionary, which could be used in a number of different trading processes, should be defined. These dictionaries, however, do not carry information about complex relationships between the elements they describe. Such information may be contained

Listing 10 Exemplary structure of the *accept-proposal* message

```
(accept-proposal
 :sender ex:Operator
 :receiver ex:Buyer
 :content (
     <m3:Offer id="ex:o12345-67" offeredPrice="-140.00"
               acceptedVolume="80.00">
      <m3:description>Buy offer</m3:description>
      <m3:offeredBy ref="ex:Buyer"/>
      <m3:volumeRange minValue="0" maxValue="100"/>
      <m3:ElementaryOffer>
       <m3:offeredCommodity shareFactor="-1"
                 ref="ex:energyOn2007-04-09T08:00:00"/>
      </m3:ElementaryOffer>
     </m3:Offer>
 )
 :language m3xml
 :ontology EnergyMarket
 :in-reply-to ex:o12345-67
)
```

in ontologies. In the body of communicative act, the ontology can be specified. The ontology clarifies the meaning of the content of the message, and also specifies its semantics. Ontologies are crucial in the case of a real e-commerce systems. This follows from the fact that the various agents can understood the same concepts in the different way, so terminology can be non-conventional, customary and confusing. More about the use of ontologies in the models of trade can be found in the chapter *A Semantic Web approach to the M^3 model*.

5 Conclusion

In the chapter two communication languages: KQML and FIPA are analyzed. The proposal for using agent communication language in the specific communication architectures considered in M^3 model is formulated. The purpose was to enable communications between the agents, while maintaining the advantages resulting from the application of this model. It was analyzed whether it is possible to adapt the FIPA standards in such a way, to allow the use of multi-commodity trading. It was also examined whether the FIPA standard covers all types of messages, considered in the M^3 model during bidding, transactions, tendering, negotiations and all related communication processes.

References

1. Austin, J.L.: How to Do Things with Words. Harvard University Press (1975)
2. Finin, T., Weber, J., Wiederhold, G., Genesereth, M., Fritzson, R., Mckay, D., McGuire, J., Pelavin, R., Shapiro, S., Beck, C.: Draft specification of the KQML agent-communication language (1993)

3. Kacprzak, P., Kaleta, M., Pałka, P., Smolira, K., Toczyłowski, E., Traczyk, T.: Communication Model for M^3 – Open Multi-commodity Market Data Model. In: Proc. 2nd National Scientific Conference on Data Processing Technologies KKNTPD 2007, pp. 139–150. Poznań (2007)
4. Kacprzak, P., Kaleta, M., Pałka, P., Smolira, K., Toczyłowski, E., Traczyk, T.: Modeling distributed multilateral markets using Multi-commodity Market Model. In: Świątek, J., Borzemski, L., Grzech, A., Wilimowska, Z. (eds.) Information Systems Architecture and Technology: Decision Making Models, pp. 15–22. OWPW, Wrocław (2007)
5. Kaleta, M., Pałka, P., Toczyłowski, E.: Multi-agent platform for trading in a distributed networks. Rynek Energii I(III), 16–22 (2009)
6. Nimis, J., Lockemann, P.C.: Robust Multi-Agent Systems The Transactional Conversation Approach. In: 1st International Workshop on Safety and Security in Multiagent Systems (SASEMAS 2004), New York (2004)
7. Pałka, P., Kaleta, M., Toczyłowski, E., Traczyk, T.: Use of the FIPA standard for M^3 – open multi-commodity market model. Studia Informatica 30, 127–140 (2009) (in Polish)
8. Searle, J.: Speech Acts. Cambridge University Press (1969)
9. Woolridge, M.: Introduction to multiagent systems. John Wiley & Sons (2001)
10. Foundation for Intelligent Physical Agents, http://fipa.org/
11. XML Path Language (XPath), http://www.w3.org/TR/xpath
12. XQuery 1.0: An XML Query Language, http://www.w3.org/TR/xquery/

Integration between Web Services and Multi-Agent Systems with Applications for Multi-commodity Markets

Dominik Ryżko and Weronika Radziszewska

Abstract. This chapter analyzes possibilities of integration between Web services and multi-agent technology. Efforts of Agents and Web Services Interoperability Working Group (AWSI WG) are described, which is focused on Web Services and FIPA (Foundation for Intelligent Physical Agents) interoperability. A hybrid architecture for conducting trade in a multi-commodity markets, which is based on multi-agent approach combined with the best practices and standards of the Service Oriented Architecture is also proposed. The architecture allows the trade to be conducted by large parties with well structured and defined offer as well as smaller entities, which operate on smaller scale and do not have resources to build full featured catalogues.

1 Introduction

In multi-commodity markets, especially those with high degree of deregulation, several challenges have to be overcome in order to perform efficient trading. Searching for new offers, composing complex products out of simpler ones, matching trade partners, negotiating prices etc. are just some of the activities to be performed. Several entities involved in the process, offering different products and services, create a heterogeneous environment in which effective communication becomes a crucial task.

Such environment requires an architecture which supports distribution of components acting on behalf of different parties, with various offers possibly described in different way. Multi-agent systems (MAS) have capabilities for building such

Dominik Ryżko
Warsaw University of Technology, Institute of Computer Science
e-mail: D.Ryzko@ii.pw.edu.pl

Weronika Radziszewska
Systems Research Institute, Polish Academy of Sciences
e-mail: Weronika.Radziszewska@ibspan.waw.pl

M. Kaleta & T. Traczyk (Eds.): Modeling Multi-commodity Trade, AISC 121, pp. 65–77.
springerlink.com   © Springer-Verlag Berlin Heidelberg 2012

architectures by introducing intelligent, autonomous and proactive agents, which exchange information in order to achieve their tasks. The drawback of using MAS is that, being a relatively young area of research, it lacks industry standards related to e-commerce.

Such standards have been developed for Service Oriented Architecture (SOA). Service Oriented Architecture is a concept of building computer systems out of a collection of services which can be called in order to perform some task. Originally the motivation for SOA was to overcome problems with integration of heterogeneous information systems and to agile development of new solutions in the enterprise. Over the time, with the fast growth of the Internet and the advent of ubiquitous environments, the services have become Web services and the standards developed for the enterprises have reached their limits. However, fresh ideas in this area, often in the form of extension of existing standards, are continuously generated, especially by communities centered around Semantic Web and ubiquitous computing. Several results, especially service registries, service discovery and composition methods etc., exist and can be reused or at least serve as an inspiration for developing new architectures.

There is a wide range of applications to the Web services not only in context of SOA. Accordingly to the authors of [3] they are considered as the future of distributed computing. The above is well specified, self descriptive and platform independent technology which also goes with an idea of encapsulation and granularity. Integrating this technology with multiagent systems could improve their utility. The authors of [2] emphasize that agents may be orchestrators for Web services and at the same time the agents could gain the ability to communicate with systems which cannot understand FIPA Agent Communication Language messages which are specific for multiagent platforms.

The reminder of the chapter is structured as follows. Section 2 presents existing results in the areas of multi-agent systems, service oriented architecture and other areas, which can be applied for the architecture envisaged in the chapter. Section 3 describes the integration between Web Service Technology and FIPA Agent Technology. In Section 4 a hybryd architecture for conducting multi-commodity trade is proposed. Finally, Section 5 summarizes the results and indicates possibile future research directions.

2   Existing Results

This section describes motivation for the chapter and discusses existing research results from the areas of service oriented architectures, Semantic Web, Web services etc., which can be adapted in order to create effective mutli-agent architecture for service discovery and composition.

2.1   Service Discovery

In a distributed heterogeneous environment, discovery of services is a complex task. In order to assist users with this task, several Service Discovery Protocols (SDPs) have been developed by academia, industry and other organizations, e.g. Intentional Naming System (INS) by MIT, Ninja from Berkeley, DEAPspace (IBM) [14]. One of the features which can be found in SDPs are templates for service naming and predefined sets of common attributes.

The UDDI (Universal Description Discovery and Integration) standard developed for storing directory of services in SOA (Service-oriented architecture), suffers from two major drawbacks. Firstly it is not suited for a distributed heterogeneous environment, and secondly it lacks semantic description. To overcome these limitations, various extensions of UDDI have been proposed. In [1] enhancement of UDDI by semantic matching of Web services has been proposed. It extends UDDI inquiry API to include specification of service capabilities. Semantic matching and automatic service composition by means of planning algorithms were also included. Services are automatically executed using BPEL4WS industry standard.

Another approach to this problem can be found in [13]. The implementation of DAML-S/UDDI Matchmaker is presented, which expands UDDI by semantic capabilities. DAML-S, which is an ontology for description of Web services, allows to discover them by capabilities and to encode their interaction protocols.

An interesting approach is proposed in [6], where EASY solution was proposed. It can be deployed on top of any existing SDP and enhance it to allow efficient, semantic, context and QoS-aware service discovery. EASY consists of EASY-L language for specification of service properties and EASY-M a corresponding set of conformance relations. In [10] a middleware platform MUSDAC is presented, which disseminates information about services across different environments. With the use of a generic service, a user can discover services located on different platforms with diverse SDPs.

2.2 Service Composition

Rao [9] compares two major approaches to solve the problem of service composition, namely based on cross-enterprise workflow and AI planning. The workflow approach can be done either statically or dynamically. In the first case, the abstract process model is built before the composition, and a query is prepared for selection of Web service to perform a particular step. Such approach allows for quick execution of standard processes, but lacks the flexibility in a dynamic environment.

In the case of dynamical composition, both process model and Web service selection are done at runtime. This is a more flexible approach, but is much more complex in terms of required computations. A combination of the static with dynamic approaches is also possible. In such a case the main abstract processes consist of abstract subprocesses, which are specified in detail in dynamic fashion.

The AI planning approach defines the composition problem as a five-tuple $< S, S_0, G, A, \Gamma >$, where S is the set of all possible states of the world, $S_0 \subset S$ denotes the initial state of the world, and $G \subset S$ denotes the goal state the planning system attempts to reach. Additionally, A is the set of actions the planner can perform in attempting to change one state to another one (the set of available Web services), and the translation relation $\Gamma \subset S \times A \times S$, defines the precondition and effects for the execution of each action (state change function for each service).

Apart from the two methods of service composition described above, other approaches have been proposed. An interesting example is conversation-driven composition [4]. Here, services are composed at run-time based on exchange of messages between the participants of the process. This requires more sophisticated conversation models, which involve more then one exchange of messages before a final service is composed.

The review of existing solutions in the area of Semantic Web and services for ubiquitous environments, shows several interesting results, which can be used in construction of architecture for multi-commodity trade. In particular, UDDI directories enhanced by semantic information can be used to register trade parties and their offer, and service composition techniques can be used to bundle related offers into more complex products.

3   Integration between Web Service Technology and FIPA Agent Technology

3.1   Is There a Need for Integration?

The Web service is a stateless request-response application. It is simple and easy to use by the systems on all platforms and made in different technologies. Because Web services are considered by some as the main element of SOA, increasing number of applications are using or integrating with them, because service oriented architecture is considered to be introduction to the new generation of Web-based business applications.

Web services are designed to be simple, stateless elements delivering limited, coherent pieces of information. The above makes them easy to use, but is also limits their application. To perform complicated tasks there is a need to interact, to send many different messages as well as react accordingly to the development of the situation. The interaction is a core element of agents and their interoperability.

The agent is an autonomous computer system able to interact with its environment, while the intelligent agent is a computer system which can behave reactively, proactively in its environment, and also interact with other agents (and other subjects) through some kind of agent-communication language.

The agent's features, behaviour and the way of communication are very well specified by the set of specifications developed by FIPA ([15, 16, 17]).

The Web services and the agents are characterized by following common features.

- Both have detailed specifications made by FIPA for agent systems and W3C for Web services.
- They both have got some kind of register for discovering and registering existing agents or Web services.

- Thanks to the detailed specification, they can communicate with any system that uses a certain communication protocols, independently from the technology or the platform.

The agent systems are not popular because of their complexity needed to perform complex tasks as for example negotiations. The FIPA specifications demand very specific types of messages, communication acts as well as the protocols being used by the agents. By the time when the above technology shall become popular, the agent systems can communicate with external systems using some kind of interpreters only. Integration with Web service technology would eliminate this limitation.

3.2 Agents and Web Services Interoperability Working Group (AWSI WG)

The IEEE FIPA Standards Committee has got the groups which contribute to the standards' development. One of its working groups is Agents and Web Services Interoperability Working Group (AWSI WG), dedicated exclusively to combine the agent technology with the Web service technology. The main aims of this group are [18]:

- Agent's interoperation with WS – the main goal is to enable the usage and communication with Web service to the agents.
- FIPA compatibility – maintaining the backward compatibility with the FIPA standards.
- Value addition to WS – enriching the Web service technology with the agents' cooperation and dialog abilities.
- Non-interference with WS standards – new propositions should not demand the introduction of any changes in existing FIPA specifications.
- Utilize Semantic Web technologies – AWSI WG considers this technology to be very promising.

The AWSI WG also proposed the mappings of standards between agent and Web service technology [18]:

- Service Description – Agent Description Ontologies can be compared to the Web Services Description Language (WSDL).
- Registration – Directory Facilitator (DF) plays the same role as Universal Description Discovery and Integration (UDDI) in Web service technology.
- Communication Protocol – Agent Communication Language (ACL) corresponds with Simple Object Access Protocol (SOAP).
- Semantic Language – defined by FIPA; FIPA-SL hasn't got the corresponding element in Web service technology.
- Interaction Schemes – FIPA Agent Interaction Protocols can be mapped with the Business Process Execution Language for Web Services (WS-BPEL) and Web Services Choreography Description Language (WS-CDL).

The AWSI group tries to connect agent and Web service technology in a way where the agent would become a reasoning engine for a service composition [2]. The technologies proposed by the AWSIG to be interesting methods of integration are:

- AgentWeb Gateway (NIIT/Comtec),
- OWL-P: OWL for Protocols and Processes (North Carolina State University),
- Deutsches Forschungszentrum fuer Kuenstliche Intelligenz (DFKI),
- JBees workflow management system (University of Otago),
- WS2JADE (Swinburne University),
- ACLs for WS communication (Universitat Politecnica de Catalunya),
- JADE WSIG – Web Service Integration Gateway (Whitestein Technologies),
- Web Service Agent Framework (Michael Maximilien, IBM Research Almaden),
- Semantic Web Services Language (SWSL + SWSO), OWL for Services (OWL-S), and the Web Services Modelling Ontology (WSMO),
- Open Cybele Agent Platform (Intelligent Automation, Inc.).

3.3   JADE Web Service Integration Gateway

The JADE Web Service Integration Gateway is an open-source addition to JADE, which allows registration of the agents which can be seen as the Web services. The above interprets the UDDI register to DF and DF to UDDI. It allows to call service by the JADE agent and to call agent functions by the Web service. The WSIG architecture is presented in Figure 1.

The main element of WSIG is JADE Gateway Agent, which is a connector between agents, DF and Web services with UDDI register on Axis Web server. The

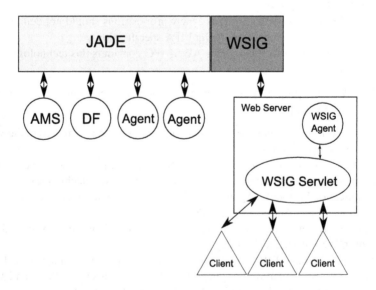

Fig. 1 Architecture of WSIG

Apache Axis is a part of "Web Services Project" responsible for communication through the SOAP. The Gateway Agent ensures that all changes in the UDDI register will also be included in the DF, so that both the agents and the Web services can simultaneously be informed about registration of any new object as well as about any changes in its characteristics. The technical details about generation of the WSDL are described in [19]. The JADE Gateway Agent also interprets the ACL messages to the SOAP format and the other way round.

WSIG requires JADE in version 3.5 or higher.

3.4   Web Services Dynamic Client (WSDC)

The WSDC [11] is an addition to JADE in version 3.7 that facilitates the invoking of the Web services without making any changes to the agent structure or calling proxy agents. It is only a library that contains all the code connected with calling agent in DynamicClient class. The programmer has only to give the URI (Uniform Resource Identifier) to the WSDL of the Web service and parameters of its call. The method *invoke* constructs a SOAP message, then sends it to the Web service and receives the response. It is not in fact a method of integration but a way to call a Web service more easily.

3.5   WS2JADE

WS2JADE [7] is a toolkit developed at the Centre of Intelligent Agent and the MultiAgent Systems. Like the WSIG, this method uses proxy agents (also one for the UDDI register) which reside in the WS2JADE system. Unlike it takes place with the WSIG, the WS2JADE creates a number of proxy agents and their relation to the Web services is many to many. This means that there would be a few agents delivering the same Web service or the Web services might be grouped and managed by one proxy agent.

The WS2JADE offers only one-way integration, it doesn't allow to publish an agent as Web service, according to [7] there is still lack of substantial theoretical work in the areas of translating agents' stateful communication to Web services' stateless one.

3.6   AgentWeb Gateway

The AgentWeb Gateway [12] enables integration of software agents and Web services without changing their current specifications. This method proposes an algorithm for efficient transformation of messages between these two technologies. AgentWeb Gateway acts as a middleware between multiagent system and Web services. It provides transformation of service discovery, service description and communication protocol.

The architecture of AgentWeb Gateway consists of search query converter (UDDI to DF converter and DF to UDDI), service description converter (DF to WSDL converter and WSDL to DF converter), service discovery converter and communication protocol converter (SOAP to ACL converter and ACL to SOAP).

3.7   SOAP MTP

The authors of [5] proposed a very interesting new approach for using SOAP in agent communication. They considered communication of two agents on different agents platform with encapsulating ACL messages into SOAP message and sending it through the net. SOAP MTP is the implementation of this idea as an add-on to the Jade agent platform. It provides a mechanism for intercepting SOAP messages, extracting ACL message and sending it to agent, of course there have to be known double addressing of agent: agent's name and a transport specific address.

3.8   Summary of Integration Technologies

In the area of integration of the multiagent systems and the Web services there is still plenty to do. However, the integration is possible and undoubted, the major problem becomes a specification and the question of finding a solution that would became the standard. There is a need for well described, stable method that would integrate both with FIPA standards and W3C. Luckily the work is in progress, the W3C in 2004 published OWL-S – OWL Web Ontology Language for Services, which reduces the gap between these two technologies.

Methods described in this document differ in ideas to challenge the problem. Some of the methods provide only one-way integration, for example the WSDC or the WS2JADE. The WSDC allows to call a service in very simple way, without any need to use any special agents, but the agents have to take care about everything. It is extremely simple solution made only to facilitate invocation of Web service by an agent.

The WSIG is designed and based on totally different point of view as it presents an agent and his functions as a Web service. It generates WSDL that includes an ontology used by that agent. Unfortunately present version of application supports one ontology only. This technology requires to describe agents in a certain way, so it requires some extra effort from the programmer. The biggest limitation is presence of only one WSIG agent in agent system, so that the reliability of the above agent is crucial for whole communication which remains in the conflict with the idea of the agents, which should be autonomous and independent one.

3.9   Example

Integration with Web services may be used to widen functionality of multiagent systems. There is a great deal of ready made Web services available through the

Internet and this is easy to use them and to reduce time needed to program a certain functionality. Moreover, many commercial systems offer their functionality through Web service by default, so any extra effort is not needed to integrate them with Web service friendly systems.

A good example of the above could be integration of commodity market with independent system, clearing house, for checking the feasibility of the transactions [8]. Agents can use a Web service of the clearing house to check if the buyer has no debts and that his offer is feasible. It would also be useful to publish agent functions through Web service: an agent system of the commodity market can have an agent working as a Web service that shares information on actual price of goods or number of agents on market to all interested parties.

Figure 4 on page 75 shows the agent system of commodity market which uses WSIG, that acts as a bridge between agent system on JADE platform and Web application on Tomcat server. Agent system can without any proxy agents execute Web service, but external systems can't "see" agents, so they need to use WSIG servlet that offers a functionality realised by agent within agent system.

4   Architecture for Multi-commodity Markets

The envisaged architecture for multi-commodity trade consists of heterogeneous domains. The main reason for such approach is that such trading platform does not start operating in the void and several legacy solutions exists. By allowing distribution of control and standards, it is possible to embed existing solutions into the new architecture, with minimal need for its reengineering. Domains can be also created on purpose to group trading parties by some specific criteria e.g.: common capital groups, use of similar standards, geographical location etc.

To allow trade to be performed across domains a fundamental layer responsible for routing of information across the system needs to be introduced. This task will be managed by special kind of agents, which will be called *backbone* agents. The routing functionality is obligatory to perform trade in the whole system, however without routing trade within particular domains is possible. All other functionalities will be build on top of routing capabilities. Figure 2 shows major functionalities and their respective relation.

On top of the routing layer, directory layer will operate. Each domain should contain one directory, which can be implemented in the form of semantically enhanced UDDI directory. The backbone agents will route queries to the remote directories and provide returned results if necessary.

Using semantically enhanced directories does not guarantee the same meaning of terminology used across different domains. Therefore a semantic disambiguation layer will be needed, which will serve as a translator between various ontologies used in the system. This layer is optional if the parties involved in trade use the same terminology.

Another optional functionality is brokering. If a party has limited resources to constantly search for attractive offers, it can use brokers to perform this task.

Fig. 2 Structure of functionalities in the hybrid
architecture

| Transaction |
| Brokering |
| Semantic |
| Directory |
| Routing |

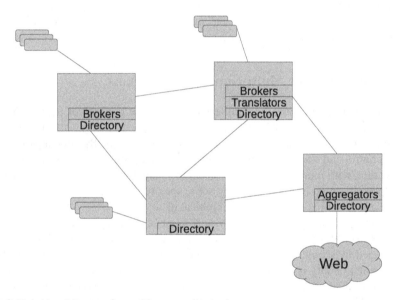

Fig. 3 Hybrid architecture for multi-commodity trade

Brokers can also implement service composition capabilities, which allows them to
seal more complex deals among multiple parties for more complex products.

Figure 3 shows an example of the envisaged architecture of the system. Trading
parties connect to selected domains and register in its directory. Other functionalities
(semantic, brokering) can be located across domains as desired. Selected domains
can act as an entry point for trading parties, which do not register proactively. Aggregator agents will be responsible for searching and extracting their offer for example
from the Web and inserting it into the directory.

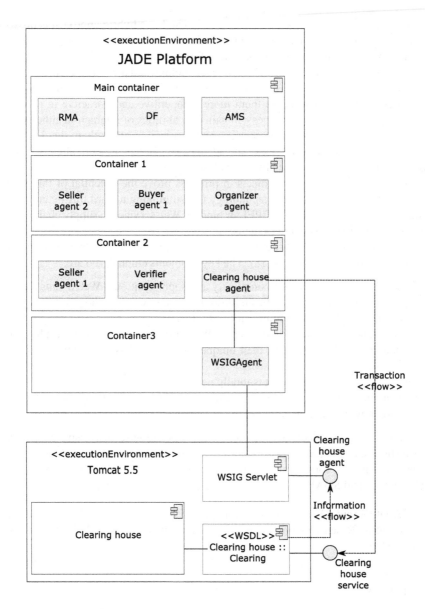

Fig. 4 Example of agent system and Web service integration

5   Conclusions

In the chapter possibilities of integrations between Web services and multi-agent technology have been analyzed. Work in the area of WS and FIPA interoperability conducted by AWSI WG has been described. An architecture for multi-commodity markets has been also proposed. It has been shown that a hybrid approach is possible, which bennefits both from the standards developed by Web service community and the flexibility and distributed character of multi-agent approach.

The integration between the Web service technology and the multi-agent systems seems beneficial for both sides. For agents systems the possibility to communicate with external systems can make them more competitive and attractive in area of business applications. Web services can gain new abilities of interaction either with the customers or with other Web services and this way to allow them the delivery of more complex business services. The agents also can benefit from already made business functionality delivered by the Web services.

The huge effort and many projects in progress prove the potential of this kind of integration. The Web service technology is also moving toward this by creating business processes, such as WS-BPEL or binding Web services to supply chains and that also can be made by agent systems.

The AWSI WG has already gathered functional and non-functional requirements for cooperation of technologies [18]. The key problems emphasized are:

- the difference in the level of complexity and sophistication – Web agents are so popular because they are small, simple and coherent, whereas the power of agent lays in its complicated communication abilities,
- the difference in addressing concept – Web service is localised by a physical endpoint address, agents have their unique ID and the physical location remains irrelevant,
- ontology – for Web services it is totally new idea,
- mathematical basis of the communication – an interaction protocol should be supported by mathematical model.

There are still some aspects not mentioned as for example security aspects of such inter-technological communication and its efficiency as well as the require of additional resources – for example increasing time of response due to interpretation between ACL and SOAP.

Acknowledgements. Partial financial support for W. Radziszewska from the Polish State Scientific Research Committee within the grant N N519 580238 is gratefully acknowledged.

References

1. Akkiraju, R., Goodwin, R., Doshi, P., Roeder, S.: A Method for Semantically Enhancing the Service Discovery Capabilities of UDDI. In: Proceedings of IJCAI 20203 Workshop on Information Integration on the Web (IIWeb (August. 2003)
2. Greenwood, D., Lyell, M., Mallya, A., Suguri, H.: The IEEE FIPA approach to integrating software agents and Web services. In: AAMAS 2007: Proceedings of the 6th International Joint Conference on Autonomous Agents and Multiagent Systems, Honolulu, Hawaii, pp. 1–7 (2007), http://doi.acm.org/10.1145/1329125.1329458 ISBN 978-81-904262-7-5

3. Liu, R., Chen, F., Yang, H., Chu, W.C., Lai, Y.B.: Agent-Based Web Services Evolution for Pervasive Computing. In: Asia-Pacific Software Engineering Conference, pp. 726–731. IEEE Computer Society (2004),
http://doi.ieeecomputersociety.org/10.1109/APSEC.2004.18
4. Maamar, Z., Mostefaoui, S.K., Mahmoud, Q.H.: Context for Personalized Web Services. In: Proceedings of the 38th Annual Hawaii international Conference on System Sciences, HICSS, vol. 07, IEEE (2005)
5. Micsik, A., Pallinger, P., Klein, A.: SOAP based Message Transport for the Jade Multiagent Platform. In: Proc. of 8th Int. Conf. on Autonomous Agents and Multiagent Systems (AAMAS 2009), Budapest, pp. 101–104. Decker, Sichman, Sierra and Castelfranchi (2009)
6. Mokhtar, S.B., Preuveneers, D., Georgantas, N., Issarny, V., Berbers, Y.: EASY: Efficient semAntic Service discovery in pervasive computing environments with QoS and context support. The Journal of Systems and Software 81, 785–808 (2008)
7. Nguyen, X.T.: Demonstration of WS2JADE. In: AAMAS 2005: Proceedings of the Fourth International Joint Conference on Autonomous Agents and Multiagent Systems, pp. 135–136. ACM, New York (2005)
8. Radziszewska, W.: Using Web Services for Integration of Multi-Agent System of the Commodity Market with External Systems (in Polish), master's thesis, Warsaw School of Information Technology (2010)
9. Rao, J., Su, X.: A survey of automated web service composition methods. In: Cardoso, J., Sheth, A.P. (eds.) SWSWPC 2004. LNCS, vol. 3387, pp. 43–54. Springer, Heidelberg (2005)
10. Raverdy, P.G., Riva, O., Chapelle, A., Chibout, R., Issarny, V.: Efficient Context-aware Service Discovery in Multi-Protocol Pervasive Environments. Mobile Data Management (2006)
11. Scagliotti, E., Caired, G.: Web services dynamic client guide (2009)
12. Shafiq, M.O., Ali, A., Ahmad, H.F., Suguri, H.: AgentWeb Gateway – a middleware for dynamic integration of Multi Agent System and Web Services Framework. In: WETICE 2005: Proceedings of the 14th IEEE International Workshops on Enabling Technologies: Infrastructure for Collaborative Enterprise, pp. 267–270 (2005)
13. Sycara, K., Paolucci, M., Ankolekar, A., Srinivasan, N.: Automated discovery, interaction and composition of Semantic Web services. In: Web Semantics: Science, Services and Agents on the World Wide Web (2003)
14. Zhu, F., Mutka, M.W., Ni, L.M.: Service Discovery in Pervasive Computing Environments. IEEE Pervasive Computing 4(4), 81–90 (2005)
15. FIPA Abstract Architecture Specification. FIPA (2002)
16. FIPA ACL Message Structure Specification. FIPA (2001)
17. FIPA Communicative Act Library Specification. FIPA (2002)
18. Agents and Web Services Interoperability Working Group: IEEE Foundation for Intelligent Physical Agents Standards Committee (FIPA SC)
19. JADE Board: Jade Web Services Integration Gateway (WSIG) Guide (2008)

1. Liu, K.; Chen, L.; Ydue, H.; Cui, W.; Li, H.; Yu, H.: Agent-Based Wireless Sensor Solution for Perishable Food Logistics. In: 5th International Conference on Engineering & Technologies, pp. 72–79. IEEE Computer Society (2009)
2. IEEE 802.15.4: Low-rate Wireless Personal Area Networks. ANSI/IEEE Std (2006)
3. Stankovic, J.; Abdelzaher, T.; Lu, C.; Sha, L.; Hou, J.C.: Real-Time Communication and Coordination in Embedded Sensor Networks. Proceedings of the IEEE 91(7), 1002–1022 (2003)

Application of the Multi-Agent Systems in the Context of the Multi-commodity Market Model M³

Piotr Pałka

Abstract. In this chapter we concern the trading platform as the system of indepen-dent software agents, which communicate with each other. Agents' interaction lead them to the exchange of commodities. Applying proper communication language ensures compliance with relevant trading mechanism. Such an approach leads us to the design and implementation of multi-agent platforms for multi-commodity trade. Implementations in Java language and in AIMMS environment are described.

1 Introduction

Multi-agent system is a system composed of two or more autonomous software agents communicating with each other and striving for their own purposes. Such a system should achieve some overarching objectives and should operate in accor-dance with intentions of the system designer. Nevertheless, the system does not implement these objectives directly, but through individual objectives of each of the agents and their interactions [21].

Such a definition of multi-agent system is closely related to the market mech-anism definition. The market mechanism should coincide to an outcome specified by the mechanism designer, under the individual strategies of particular participants [12]. It is particularly important in multi-agent exchange platforms, where individ-ual agents can act strategically in order to draw some unjustifiable advantage of the market.

Multi-commodity trading concerns widely recognized infrastructure markets, with complex infrastructure and security constraints, e.g. electrical energy markets [19], bandwidth trading in teleinformatics networks [4, 18], allocation of railway re-sources [3], greenhouse gases emission trading markets [5, 14, 17]. It not only makes use of bundles of commodities, but also allows for complex bidding processes. Also,

Piotr Pałka
Warsaw University of Technology, Institute of Control and Computation Engineering
e-mail: P.Palka@ia.pw.edu.pl

M. Kaleta & T. Traczyk (Eds.): Modeling Multi-commodity Trade, AISC 121, pp. 79–98.
springerlink.com   © Springer-Verlag Berlin Heidelberg 2012

both centralized trading on the exchange platform, as well as bi- or multilateral negotiating contracts on the distributed market may occur. The Multi-commodity Market Model (abbrev. M^3) [23] is a proper data model for many complex markets, and it can be used for description of the multi-agent platform problems. Therefore, we use M^3 to model market relationships in the multi-agent platform for multi-commodity exchange (see Fig. 1).

2 Multi-Agent Exchange Platforms

The market system can be modeled as a game between market participants (see Fig. 2).From implementation point of view, individual market participant can be considered as an autonomous entity, guided by its own interests. On the Fig. 3 we can see particular market entities (e.g. the generating stations), which are represented by agents. The individual market participant comes to interact with other participants, motivated by an intention to achieve certain gains from the exchange of goods. Market institutions (e.g. centralized exchange market or negotiated contracts platform) allow participants to trade by establishing certain rules under which trade takes place. e-Bay online auction can be an example of a market institution, where the trade is performed according to the second price, one-sided auction mechanism.

To obtain a harmonization of individual goals of every market participant with global goals desired by the mechanism designer, a use of proper mechanism should be considered.

Trading processes on different markets (e.g. electrical energy markets), related to negotiations and bidding, are specific for each type of trade. Multi-agent systems are natural framework for automating the task of the trading processes in the environment, where many entities have their own objectives. In the chapter we consider a concept of a multi-agent exchange platform for trading on various markets. In the multi-agent system, an agent is a software component which is able to take autonomous actions and conduct negotiations, to submit offers and exercise sophisticated strategies. However, trading processes can be performed in different circumstances and may take different forms. For instance, they can be related to bi- or

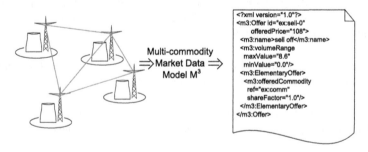

Fig. 1 Infrastructure model notated as a M3-XML document

Fig. 2 Market system modeled as a game between participants. Particular agents
(B1, G1 and G2) are as-
sociated to network nodes.
Network infrastructure creates additional constraints.
System operator shall en-
sure compliance with these
additional constraints.

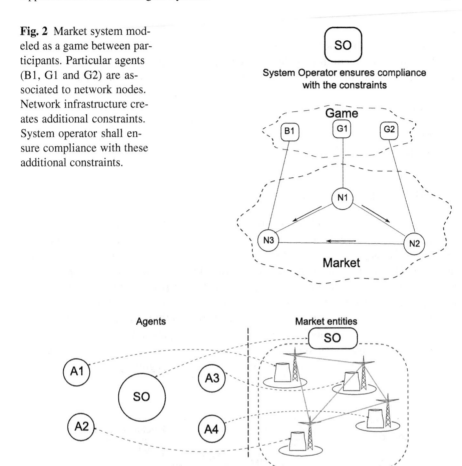

Fig. 3 Market entities represented by agents in the multi-agent system

multilateral negotiations in distributed environment, or may have more formalized
conditions imposed by some regular exchange [10].

We assume, that developed multi-agent trading platform should comply with different market architectures. Detailed description of such market architectures is contained in chapter *Communication models used in the context of multi-commodity
trade*.

3   Design of the Multi-Agent Platform for Multi-commodity
Exchange

In this section we present a design of the multi-agent trading platform. Proposed
platform should include the following elements [16]:

- agents that represent actual market participants;
- multi-agent environment, in which these agents will be embedded; it should implement a way of communication – agents communication language;
- a bidding model that allows for a flexible submission of asks and bids;
- a market mechanism, which analyses signals obtained from agents, performs allocation of the commodities, and sets prices.

Diagram of designed multi-agent platform is depicted in Fig. 4. Following sections present description of the platform elements.

3.1 Structure of the Agents

Every agent, as set forth in [8], should implement basic trading agents roles, or the aggregating or intermediating agents roles. The agent can also implement a number of roles, and change them, depending on the situation, e.g. while communicates with the aggregated agents, the agent plays an aggregating role, and while it is negotiating with other agents, it plays the negotiator role. Thus, the participation of every agent in widely understood multi-commodity trade is possible. For example, particular agent can agree a contract on selling a bundle of commodities (for the sake of clarity, let it be the A and B commodities). However, the agent may not currently posses the B commodity. Thus, it can simultaneously announce a new auction for buying the B commodity, so the agent can play simultaneously the negotiator and the operator role.

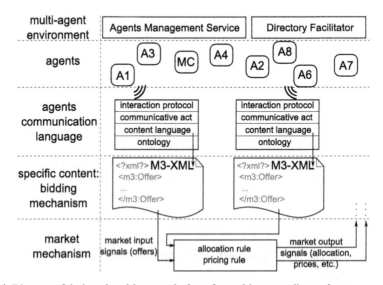

Fig. 4 Diagram of designed multi-agent platform for multi-commodity exchange

3.1.1   Basic Trading Agent Roles

Negotiator Agent

Negotiator agent carries out the best, from the real negotiator point of view, contracts. Agreed contracts are preceded by (possibly complex) negotiations. Negotiating agents have wide freedom in applying negotiation strategies. Agent can simultaneously perform negotiations with more than one agent (see Fig. 5). Moreover, we assume, that agents can perform multilateral negotiations.

Trading Agents

Trading agents are divided into two main groups: sellers and buyers. Seller-agent is an intermediary between the real seller and the multi-agent system. Similarly, the buyer-agent is an intermediary between the real buyer and the system. Trading agents submit asks or bids for desired commodities, motivated by a desire to directly realize the goals of physical market participants (see Fig. 6). A single trading agent, in a case of multi-commodity trading, can trade with bundles of commodities by submitting single bundled offer. The commodities, according to multi-commodity market model, can be sold, bought or exchanged.

Operator Agent

Operator agent carries out balancing of asks and bids, between members of some group, which it represents. Operator institution allows members of such group to preserve anonymity during transaction execution. Simultaneously it provides important market signals, e.g. prices. Operator-agent acts as a central entity, which collects asks and bids of other agents, performs offers allocation, sets prices, and sends the results to the trading agents (see Fig. 6).

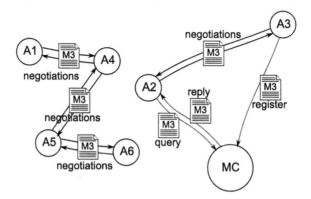

Fig. 5 Negotiating agents in the multi-agent environment

3.1.2 Aggregating and Intermediating Agents

In multi-commodity centralized or distributed trading platforms, also more complex agent roles may appear (e.g. intermediator, broker, aggregator, etc.) [10]. During gradual market evolution, companies intermediating in turnover emerge in a natural way. Such companies become exchange and trading operators. Every trading operator represents a group of agents in trading. Therefore, except for basic agents (i.e. trading and negotiating agents), we can distinguish aggregating and intermediating agents on the market [20].

Intermediator Agent

Intermediator agent provides business and services on behalf of agents represented by it. Particularly, it prepares asks or bids, according to preferences of agents it represents. Afterwards, it submits those bids on specified turnover platforms, where it often negotiates the terms of trade, again, on behalf of other agents.

Reseller Agent

Reseller agent takes the risk relevant to the trade. Reseller signs the contracts on buying the commodities, with its owners or providers. Afterwards it resells previously bought commodities to prospective buyers. The economies of scale affects the profitability of contracts.

Aggregator Agent

Aggregator agent represents the group of common suppliers and customers in an external offer processes. Particularly, it prepares asks and bids on the relevant commodities and services, taking into account interests represented by agents it represents. Afterwards it submits those bids on determined turnover platforms, on behalf of such agents. Such offers could be aggregated; the aggregation concerns agents,

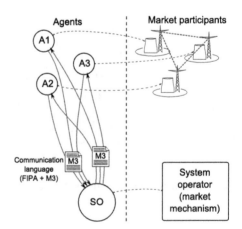

Fig. 6 Schema of agents' interaction on the centralized market and the dependencies between agents and market participants

not commodities. Thus, aggregated offers are aggregations of particular offers, concerning particular commodities.

3.1.3   Morris Column Agent

A particular agent, to be able to negotiate, needs to collect some specific information about other agents, and particularly about the current trade process. We can certainly imagine a case in which each agent broadcasts the messages about initiating of some new trade to all other agents. However, in large systems, such a solution would involve a massive communication effort. Moreover, agents would obtain a huge amount of unnecessary information.

Thus the Morris Column agent role was proposed [8]. The task of such agent is to offer some public location, where other agents may "hang a notice" to report certain information about trade processes provided by them. In addition, the Morris Column agent should provide a functionality of searching and removing the information. In a broader context, agents may leave notices on initiating various types of auction processes with some market-specific characteristics.

Both in distributed and centralized market systems, more than one agent can exist, who plays a Morris Column role. However, in a distributed market system, there is a strong need for multiple instances of such agents. Every single Morris Column agent may store only a part of global information, and particular pairs of such agents may share some parts of information.

3.2   Multi-Agent Environment

Agents implemented as software components need to be embedded in a certain environment. Standards, delivering solutions, in which agents can co-exist and interact, are defined by FIPA (Foundation of Intelligent Physical Agents) [22]. FIPA, as an organization that promotes use of multi-agent environments, has developed several standards for multi-agent communication, has clarified necessary Communicative Acts, an has identified and proposed a number of message exchange protocols. The main frame is the Agent Management Reference model. In this model, every agent is embedded in a standardized platform, which provides the main system for agents communication. On this platform, the agent must operate with awarded AMS (Agent Management Service) that manages a registry of all agents acting within the platform. In addition, the Directory Facilitator (DF) agent can be defined, which informs agents about existence of other agents and services they provide.

Despite apparent compliance with the Morris Column agent, Directory Facilitator Agent operates in limited way. It accepts queries as a simple strings, interpreted by him as the names of the services registered by other agents, and returns only identifiers of matched agents. More detailed description of the limitations in implementing the Morris Column agent as the extension of Directory Facilitator agent can be found in section 4.3.

3.3   Agents Communication Language

Detailed description of agents communication language is contained in chapter *Communication models used in the context of multi-commodity trade*.

3.4   Bidding Mechanism

Agents should have the flexibility to express their preferences by means of the parameterized offer. As a part of a broad spectrum of organized trading platforms, an offer may take various forms. Offer's form is evolving with the development of market solutions, awareness and the requirements of the participants. Therefore, it is essential that the system should be able to express a wide range of possible forms of offers. This objective can be achieved by using the expressive offer model proposed by the M^3, consisted of three main groups of offers.

The simplest offer type, an elementary offer, is an offer by which the offerer can express his preference on a single commodity, including the unit price and the maximal allowed trading volume.

The second offer type is a bundled offer. This is a characteristic type of offer for multi-commodity turnover, where players trade with packages (or bundles) of commodities with fixed proportions of commodities in the bundle.

The most complex type of offers is a grouping offer. It aggregates a set of other elementary or bundled offers, and describes relationships between these offers. There is no collective offer price, but the relationships described by the grouping offer must be satisfied after balancing. However, as a result of balancing, some cost can be assigned to such offer. Grouping offers allow the market entities to define individual constraints. Fulfillment of a single constraint can be treated as an individual service. For example, power generation unit needs to be started-up before power production, thus an individual start-up service must be provided. Of course, grouping offers can be used to describe much more complex relationships.

More formal model of foregoing offer types is presented in chapter M^3 – *motivations and formal model*.

3.5   Market Mechanism

The mechanism theory is a science, which determines how to design principles of games (or markets, which can also be treated as a class of games) to get a result with the relevant properties. Each agent participating in the mechanism is interested in maximizing its individual utility function.

The mechanism, from implementation point of view, can be regarded as a programming interface, which ensures a compliance with certain desired properties if they are implemented in the appropriate system. Multi-agent systems can implement such interface. Properly designed mechanism should ensure the harmonization of the individual agents' objectives (e.g. payment) with the global objectives (e.g. economic welfare). It is particularly important in multi-agent trading platforms, where individual agents can act strategically in order to draw some unjustifiable advantage of the market.

The mechanism theory provides clearly defined mechanisms properties. These are desirable properties, and each properly designed market mechanism should strive to meet them all. We can use these properties as a certain criteria, which will provide us an information about the "quality" of given mechanism. Mechanism designer, having knowledge of the market, including subjective structure, resources, and network infrastructure, in which the trade takes place, can identify the most desirable properties of the market mechanism.

In the previous subsection, we consider the M^3 model application to the multi-agent trading platform. Thanks to the M^3 model we can implement various market mechanisms, including various allocation rules and various pricing rules. Therefore, we propose a multi-agent trading platform, which can handle with various mechanisms. Thus, we obtain a general-purpose trading platform, which can be used in a wide class of applications. As set forth in [9], the relevant mechanisms can be implemented using the M^3 model.

4 Java Implementation

In this section we describe the implementation of a multi-agent trading platform in Java language. We chose to base the implementation on Java language for the following reasons. Java provides the XML binding model (JAXB) [13], which allows separating the data format from the implementation of the system. Second reason is an existence of recognized FIPA standards implementation in Java-based JADE framework [1]. The third reason is an availability of the decisional-computational processor in Java language [9]. Such processor can be used for implementing mathematical models of market problems (see Fig. 7).

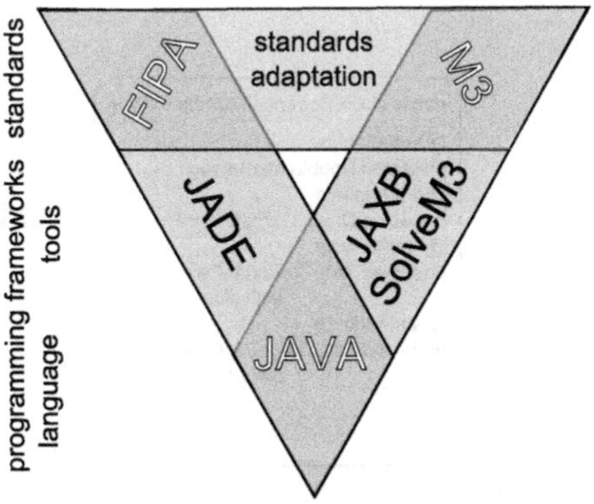

Fig. 7 Standards, frameworks, tools and programing language used in implementation the multi-agent trading platform

4.1 Used Technologies

During development of multi-agent trading platform, the following frameworks were used.

Java Agent Development Framework (abbrev. JADE) system is a practical implementation of standards proposed by FIPA [1]. It provides skeletal solutions for the architecture and defines ways of interaction between agents. It provides developers with a runtime environment, where agents can be embedded, and they can communicate with each other from various network locations. Also, JADE provides user with the intuitive library of classes, which enables developing of agents, their behavior, and the message exchange patterns.

Java Architecture for XML Binding (abbrev. JAXB) provides XML documents binding into Java classes [13]. Such binding facilitates access to XML data without necessity of knowing the complex XML document structure. It is mainly used for processing the M3-XML notation – XML based implementation of M^3 model (see Fig. 8).

The decisional-computational processor [9] can be used as a universal framework for solvers for optimization of market models (in following part of this chapter we will use shorter name – SolveM3). The M^3 model allows to formulate a set of market data, which may be easily exchanged or/and shared between users of various market clearing systems. Thus, decisional-computational tool, which should ensure proper offers matching and price setting, is needed. The decisions should be taken basing on the data expressed in M3-XML notation. The results obtained by those tools should also be written in M3-XML notation (see Fig. 9).

4.2 Implementation Details

In this section we describe details of the implementation of the multi-agent platform. The agents are implemented as the Java threads, and strategies of agents are implemented as Java methods. Agents, to be able to communicate, are embedded in JADE environment, which provides them with communication methods.

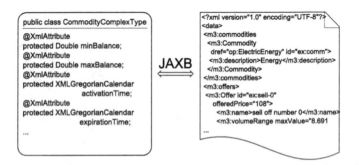

Fig. 8 Marshaling and unmarshaling the M3-XML file into Java classes using JAXB

In order to facilitate the use of XML content, the JAXB technology is used. Through this technology, the XML structure is mapped into Java classes, which facilitates the use of XML data in Java code. Also, after processing of the data within the code, the structure of classes is transformed back into XML.

When a centralized trade model is considered, various market mechanisms can be used to obtain proper commodities reallocation and pricing according to the established rules. Market mechanisms take into account constraints specific for a given problem, e.g. system, infrastructure or security constraints. Market mechanisms are implemented as the decisional-computational processor [9], which takes input data in M3-XML format, solves the market problem, and returns results also in M3-XML format. A specification of a market problem is translated to the form accepted by the computational processor, by specifying an appropriate XSLT transformation. In the case of decentralized trade, i.e. negotiated contract platform, the M3-XML notation is also used. The reason is to add more flexibility in expressing agents preferences by means of the M^3 model.

The multi-agent exchange platform uses XML language (mostly M3-XML dialect) on both the message content and the market modeling layers.

4.3 Encountered Problems

During the implementation of the multi-agent platform for multi-commodity exchange, some problems were encountered. We intended to implement the Morris Column agent as the Directory Facilitator (abbrev. DF) agent's extension [1]. As set forth in [11], the Morris Column agent should exchange messages with complex XML content. The Directory Facilitator provides other agents with information about the existence of other agents and services they provide. Unfortunately, the DF agent implementation is limited in the context of message content possible to send – it does not provide the standard message content, thus XML, XPath or XQuery content, which is required by the Morris Column agent (see [11]), cannot be used.

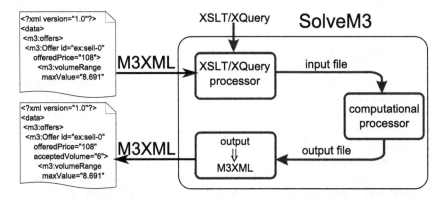

Fig. 9 SolveM3 application overview

Therefore, the Morris Column agent cannot be implemented as an extension of the Directory Facilitator agent, and must be implemented as a regular agent extension.

As set forth in [10, 15], some of JADE interaction protocols have to be modified, to enable better use of the M^3 capabilities: multi-commodity trading and complex, multilateral negotiations. JADE enables application of the interaction protocols proposed by FIPA, nevertheless it does not provide its extension. This is disadvantage of the JADE framework. In the [6] authors describe the implementation of new interaction protocols, using the events, a multi-level tree of dispatchers that match and route events, and a hierarchical state machine that is based on the UML state chart model. The extension of the *Contract Net Interaction Protocol* and the *Iterated Contract Net Interaction Protocol* was implemented using the hierarchical state machine technology.

4.4 Java Implementation Multi-Agent Platform Testing

A simulation of the multi-agent platform for multi-commodity exchange was performed on ten computers, connected with the 100 Mb/s LAN network. Each computer, with Intel Core 2 Duo CPU E8400 3.00GHz processor and 3 GB RAM, runs Ubuntu 8.04 64-bit. On every computer the multi-agent platform for multi-commodity exchange works on JRE version 1.6.0_17, with JAXB framework as the XJC plugin for Eclipse 3.1.0+, based on JAXB 2.1 Final Release, and JADE framework version 3.7.

We perform four test cases, which differs in agents location, and the network configuration (see Fig. 10).

Number of agents participated in the simulation: one agent plays the Morris Column role and the rest of them have the negotiator roles. We assumed, that half of negotiators reported willingness of negotiating to the Morris Column agent, the other half chose one partner from the Morris Column in random way, and started the negotiations. We assumed also, that particular negotiator agent conducted negotiations on behalf of actual market participants (producers, distribution companies, customers, etc.). In this test study, we assumed one traded commodity, which is an electrical energy.

4.4.1 Negotiation Results

Agents came to a solution through a simple negotiation strategy. The strategy assumes that the chosen agent begins negotiations from a fixed price. Agent receiving the proposal decides if it is acceptable. If so, it responds with a new price, constantly approaching to its reservation price. Otherwise, the agent breaks the negotiations. After a number of proposal exchanges, agents came to an agreement.

Typical negotiation process is as follows. Agent A1 plays the Morris Column role, agent A3 is willing to sell 75 [MWh] of energy, its reservation price is equal to 35 [$/MWh]. It registers willingness to sell on the Morris Column. Agent A2 is willing to buy 100 [MWh] of energy. Its reservation price is equal to 66 [$/MWh]. Agent A2 queries the Morris Column, and founds the negotiation partner: A3. A2 proposes purchase of 100 [MWh] energy for 33 [$/MWh] to A3. Agent A3 responds with a proposal for selling 75 [MWh] energy for 70 [$/MWh]. Agent A2 modifies its proposal: it proposes purchasing 75 [MWh] energy for 43 [$/MWh]. Agents came iteratively to the solution. Finally the contract on 75 [MWh] for 53 [$/MWh] between A2 and A3 is agreed. As we can see, agents came to the acceptable solution.

4.4.2   Technical Properties

In this section we analyze technical properties of the platform, namely elapsed times for communicative acts and interaction protocol execution.

An average elapsed time for sending single communicative act (in this case it is the *propose* act) from one agent to another, is measured. The time is measured from the beginning of sending the message to the end of receiving it by the receiver agent. Because of the short time for single communicative act exchange between two agents, the experiment is performed by exchanging the same communicative act 10 000 times, between pairs of agents chosen randomly. Results are presented in Table 1. Notice, that with increasing number of agents, the communicative act exchange was performed simultaneously between every pair of agents.

An execution time of communicative acts exchange between two agents, according to specific interaction protocol, was measured. Marshaling and unmarshaling

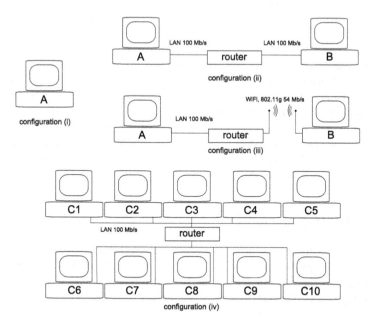

Fig. 10 Experiment configurations. Case (i): all agents are running on the single machine. Case (ii): agents are running on the computers A and B, computers are connected with the router by wired connection. Case (iii): agents are running on two computers: A, connected to router by wired connection, and computer B is connected to router by wireless connection

Table 1 Aggregated results for single communicative act (*propose* act) exchange

No. of agents	Avg. time [ms]	Std. dev. [ms]	Min. time [ms]	Max. time [ms]
100	1.14	0.38	0.17	1.87
500	7.74	1.77	2.19	10.78
1 000	18.48	7.17	5.38	31.72

the XML data, and sending as well as receiving particular communicative acts are included. In this case, we analyze the modified *Iterated Contract Net Interaction Protocol*, as the interaction protocol applied during negotiations. Notice, that the negotiations can fail, and it does not affect the assessment. Results are presented in Table 2.

Table 2 Aggregated results for modified *Iterated Contract Net Interaction Protocol* duration

No. of agents	Avg. time [ms]	Std. dev. [ms]	Min. time [ms]	Max. time [ms]
100	103.23	102.35	20	387
500	56.59	49.57	7	293
1 000	71.06	70.39	7	493
5 000	189.69	237.71	12	1 486
10 000	167.57	290.18	9	3 658

In the experimental study we show that multi-agent platform for multi-commodity exchange enables individual agents to achieve reasonable results through their interactions. Moreover, analyzed exchange single communicative act and negotiation times show that the multi-agent platform can be an effective tool for implementing negotiated contract platforms. Results of simple experimental study show, that in the proposed solution, either elapsed time for single communicative act exchange, as well as elapsed time for interaction protocol, in various network configurations, are relatively short.

5 AIMMS Implementation

This section presents another implementation of the M^3 model in an open platform for multi-commodity exchange. It is implemented as a multi-agent system in which individual agents represent individual market entities, or their groups. The platform has been implemented in AIMMS environment.

5.1 Used Technologies

The Java implementation of the multi-agent platform for the multi-commodity exchange, has several drawbacks, especially when it comes to introducing large chunks of data, and also in the case of creating the new, complicated market mechanisms. A good tool to solve these problems is the AIMMS (Advanced Integrated Multi-dimensional Modeling Software) [2] environment. AIMMS provides a set of optimization tools that make it possible to implement the balancing of the various market segments in the form of mathematical optimization models. AIMMS also offers support for building multi-agent environments. AIMMS also has the possibility of creating the multi-agent system, moreover it provides tools to facilitate import and export data stored in XML notation. This makes it possible to load data stored using an M3-XML dialect into the AIMMS and to export the AIMMS data to a format compatible with the M3-XML.

AIMMS enables flexible implementation of the multi-commodity market mechanisms, allowing the integration of optimization package, a flexible modeling language and a high-level programing language operating directly on the elements and data of optimization problems. AIMMS also includes integrated tools for creating user interface for the presentation of resources, e.g communications network in the form of a visual representation of the graphs, or to the visual presentation of the electricity transmission network. Ability to integrate these elements in multi-agent environment makes AIMMS good solution for the implementation of multi-agent platform for the multi-commodity exchange (see Fig. 11).

In AIMMS, the rules of the mechanisms are formulated as the optimization problems. Complex problems of balancing market can be solved through sequence of elementary tasks of balancing realized in the internal AIMMS language. This is very laborious, but also provides a potentially high degree of flexibility, i.e. a broad spectrum of market problems can be implemented. Each of them (e.g. the electricity market, the market for telecommunication resources etc.), has some specific operating conditions.

5.2 AIMMS Multi-Agent Environment

Agents are implemented as software components, and therefore must be embedded in a multi-agent environment, which will enable them to communicate. AIMMS offers such multi-agent environment.

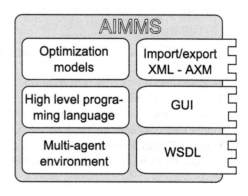

Fig. 11 Elements of
AIMMS environment

In this environment, a so-called AIMMS message queue (identified by a string) is created. The messages sent by agents are directed to the message queue. The message queue send messages to recipients. With such a construction, it is possible to implement various market organizations. We can also imagine a case when a single agent connects to a single message queue and then to another. Such scenarios have been considered in [7].

Different types of agents in the environment correspond with the AIMMS agent roles. AIMMS unfortunately does not implement the FIPA standards. However, it enables defining message types, and implementing interaction protocols, in a manner consistent with the FIPA proposals. Individual acts of FIPA communication standards are mapped to the message types in AIMMS. AIMMS forces assigning the send and receive of individual message operations to individual agents, in the form of particular procedures. After this configuration, the AIMMS automatically creates skeletal methods of sending and receiving all the messages for each agent. AIMMS provides object-oriented programming language, which enables to implement specific agents behavior, and individual agents' strategy.

An important feature is the possibility to issue a WSDL interface (Web Services Definition Language) and provide the desired functionality through the Web Services mechanism. Due to this, the cooperation between the AIMMS and the agents implemented in other programming languages is possible (see also chapter *Integration between Web services and multi-agent systems. . .*).

5.3   Communication Language

The AIMMS does not meet the FIPA standards. Thus the implementation of communication acts as the internal procedures in object-oriented programming language was enforced. We implemented the communication between the operator agent and the trading agent basing on the *Contract Net Interaction Protocol* [22].

The content of the AIMMS message is composed of a number of fields. Each field is assigned a type (e.g. number, string, set element). In the implementation we assume that message content will be interpreted as the M3-XML document. With the conversion mechanism described in section 5.4, it is possible to load data from M3-MXL document to the AIMMS data structures. In this way we get the flexibility of the multi-agent platform for the multi-commodity exchange.

5.4   The M^3 and AIMMS Integration

We prepare special procedures in AIMMS, to operate on the M3-XML documents. AIMMS can read and write data from/to XML files. This is made possible by mapping individual variables, parameters and elements (AIMMS data structure) on individual elements and attributes of the XML file. This mapping is stored in an AXM file (AIMMS Mapping XML file). Based on the schema definition files XSD (XML Schema Definition), using the built-in XML Schema mapping tool, one can create an AXM mapping file. AXM files were generated for both write and read. To process input and output M3-XML files we use XSLT transformations (see Fig. 12).

5.5 Implementation Details

To implement multi-agent trading platform, we use AIMMS RPC (Remote Procedure Call) module. Implementation concerns of three areas. These areas are: the multi-agent environment configuration (Community Setup), the roles of agents configuration (Agent Setup), and the implementation of market mechanism.

We setup multi-agent environment by creating two types of agents: trader and operator. Then we define the required message types (basing on FIPA communicative acts). Next we assign the agent-message pairs, and implemented the message flow diagrams (to achieve an approximation of the Contract Net Interaction Protocol). Next, the AIMMS generates a skeleton of procedures that send and receive individual messages. We implement strategies for particular agents. Finally, we implement a set of trading mechanisms.

5.6 Capacities of the AIMMS Multi-Agent Environment

AIMMS is not a solution that provides flexible implementation of multi-agent systems. MAS built in AIMMS is not a scalable solution, mainly in developing communication. This is due to implementation of communication processes as procedures. This forces the need for rebuild the skeletons on the agents' procedures, that participate in the message exchange. This limits the possibilities of extending the multi-agent system and causes difficulties in maintenance, especially when introducing new agent roles, new messages, and changing the message content. However, when we assume that message content is written in the M3-XML dialect, this restriction becomes less important because of the flexibility of shifting the burden of the communication layer to data model layer. Another lack of AIMMS is a need

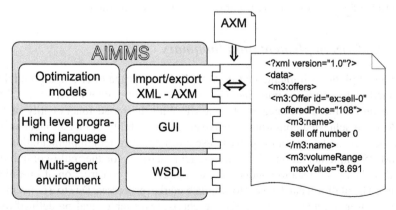

Fig. 12 Import/export M3XML data into AIMMS data structure

for implementing additional software functionality for handling message exchange patterns.

Using the AIMMS as a multi-agent environment, is supported by the possibility of connecting the remote agents to the environment. The connection is based on the host address and the message queue identifier (which can be regarded as a multi-agent system identifier). There exists the possibility of both distributed and centralized multi-agent system implementation.

Implementation of a specific type of multi-agent system, which is the multi-agent platform for multi-commodity exchange, makes it possible to use M^3 dialect M3-XML as the content of the messages (the content language). In this way we obtain the relative flexibility of the system. Moreover, the implementation of multi-agent platform in the AIMMS supports ease of implementation of market mechanisms by AIMMS optimization tools, as well as ease of implementation of the user interface.

6 Summary

The design of the multi-agent platform for the multi-commodity exchange uses world-wide standards (FIPA) and an advanced model of the multi-commodity market (M^3). It is also based on our team's experience in designing complex market mechanisms.

6.1 Java Implementation Summary

Multi-agent platform for multi-commodity exchange consists of two main modules. One is a multi-agent system, the second is the implementation a variety of market mechanisms, developed in the global research trend. Due to application of the JADE and JAXB frameworks, combined with the M^3 standard, and with the decisional-computational processor, which allows for high flexibility in modeling of mechanisms and market processes, we obtain a flexible application, which allows us to perform various market simulations, from complex, multilateral negotiations to centralized multi-commodity trading with complex constraints.

6.2 AIMMS Implementation Summary

Multi-agent platform implemented in AIMMS is composed of two main modules. First module responds for implementation of a series of market mechanisms. Second module is the multi-agent system itself. With the broad functionality of the AIMMS, we achieve a compromise between high flexibility of market mechanism and processes modeling, and slightly worse multi-agent system implementation.

AIMMS implementation of the multi-agent platform for the multi-commodity exchange, is a good idea to conduct research and for the construction of prototype systems. The platform can be used to simulate market processes. Due to a convenient graphical user interface is easy to create charts, pivot tables and predefined user windows – it makes data analysis much easier. The platform can be used to compare and design of new market mechanisms.

6.3 Overall Summary

The multi-agent platform for multi-commodity exchange can be used in a wide range of applications. Both described platforms take advantage of general Multi-commodity Market Model M^3, which enables this solution to be used for simulating market processes in a wide range of infrastructure markets.

References

1. Bellifemine, F., Poggi, A., Rimassa, G.: Developing multi-agent systems with JADE. Springer (2001)
2. Bisschop, J., Roelofs, M.: Aimms – Language Reference. Paragon Decision Technology (2006)
3. Borndörfer, R., Grötschel, M., Lukac, S., Mitusch, K., Schlechte, T., Schultz, S., Tanner, A.: An auctioning approach to railway slot allocation. ZIB Technical Report ZR-05-45 (2005)
4. Courcoubetis, C., Weber, R., Coe, M.: Pricing Communication Networks: Economics, Technology and Modelling. John Willey & Sons (2003)
5. Ermoliev, Y., Michalevich, M., Nentjes, A.: Markets for tradeable emission and ambient permits: A dynamic approach. Environmental and Research Economics 15, 39–56 (2000)
6. Griss, M.L., Fonseca, S., Cowan, D., Kessler, R.: SmartAgent: Extending the JADE Agent Behavior Model. HP Laboratories Technical Report HPL-2002-18 (2002)
7. Kacprzak, P., Kaleta, M., Pałka, P., Smolira, K., Toczyłowski, E., Traczyk, T.: Communication Model for M^3 – Open Multi-commodity Market Data Model. In: Proc. 2nd National Scientific Conference on Data Processing Technologies KKNTPD 2007, pp. 139–150. Poznań (2007)
8. Kacprzak, P., Kaleta, M., Pałka, P., Smolira, K., Toczyłowski, E., Traczyk, T.: Modeling distributed multilateral markets using Multi-commodity Market Model. In: Świątek, J., Borzemski, L., Grzech, A., Wilimowska, Z. (eds.) Information Systems Architecture and Technology: Decision Making Models, pp. 15–22. OWPW, Wrocław (2007)
9. Kacprzak, P., Kaleta, M., Pałka, P., Smolira, K., Toczyłowski, E., Traczyk, T.: Decisional-computational processor for market data model M^3. In: Kozielski, S., Małysiak, B., Kasprowski, P., Mrozek, D. (eds.) Databases - Development Methods and Technology: Architecture, formal methods and advanced data analysis, WKiŁ, vol. 1, pp. 215–226 (2008) (in Polish)
10. Kaleta, M., Pałka, P., Toczyłowski, E.: Multi-agent platform for trading in a distributed networks. Rynek Energii I(III), 16–22 (2009) (in Polish)
11. Kaleta, M., Pałka, P., Toczyłowski, E., Traczyk, T.: Electronic Trading on Electricity Markets within a Multi-agent Framework. In: Nguyen, N.T., Kowalczyk, R., Chen, S.-M. (eds.) ICCCI 2009. LNCS, vol. 5796, pp. 788–799. Springer, Heidelberg (2009)
12. Krishna, V.: Auction Theory. Academic Press (2002)
13. McLaughlin, B.: Java and XML data binding. O'Reilly & Associates Inc. (2002)
14. Nahorski, Z., Stańczak, J., Pałka, P.: Multi-agent approach to simulation of the greenhouse gases emission permits market. In: 3rd International Workshop on Uncertainty in Greenhouse Gas Inventories, pp. 183-194. Lviv Polytechnic National University (2010)

15. Pałka, P., Kaleta, M., Toczyłowski, E., Traczyk, T.: Use of the FIPA standard for M^3 – open multi-commodity market model. Studia Informatica 30, 127–140 (2009) (in Polish)
16. Pałka, P., Całka, M., Kaleta, M., Toczyłowski, E., Traczyk, T.: Design and java implementation of the multi-agent platform for multi-commodity exchange. In: Proc. 3rd National Scientific Conference on Data Processing Technologies KKNTPD 2010, pp. 184–196. Poznań (2010)
17. Stańczak, J., Bartoszczuk, P.: CO_2 emission trading model with trading prices. Climatic Change: Benefits of Dealing with Uncertainty in Greenhouse Gas Inventories 103(1), 291–301 (2010)
18. Stańczuk, W., Lubacz, J., Toczyłowski, E.: Trading links and paths on a communication bandwidth markets. Journal of Universal Computer Science 14(5), 642–652 (2008)
19. Stoft, S.: Power System Economics: Designing Markets for Electricity. Wiley Interscience (2002)
20. Toczyłowski, E.: Optimization of Market Processes under Constraints, II extended edition. EXIT Academic Publishing (2003) (in Polish)
21. Woolridge, M.: Introduction to multiagent systems. John Wiley & Sons (2001)
22. Foundation for intelligent physical agents, http://www.fipa.org/
23. M^3 – Multicommodity Market Data Model, http://www.openm3.org/

A Semantic Web Approach to the M³ Model

Przemysław Więch and Tomasz Gidlewski

Abstract. In the constantly expanding Semantic Web, the domain of describing offers and concepts related to them plays an important role. The use of ontologies provides a means for automated processing of knowledge, which is expressed in a standardized representation language. We argue that the M³ model can be represented in OWL, a standard ontology language. We show the benefits of such a representation. Several ontologies are shown, which can complement the descriptions made in M³. We propose a translation of the M³ model and the M3-XML data format into an OWL-based ontology. We also show exemplary reasoning tasks, which could be accomplished with standard description logic reasoners. Some problems with using ontologies for expressing offers are also described.

1 Introduction

The current World Wide Web can only be viewed as a syntactic structure of Web pages, and search engines analyse only sequences of words detached from their meaning. The idea, named Semantic Web, for transforming the existing Web, envisions the knowledge contained in the Web to be expressed in a formalised way in order to make it possible to be processed on the level of semantics.

This vision applies very well to describing offers on the Internet for business to client and business to business purposes. Currently, in electronic commerce, offers are mostly described using the natural language, which can be searched for keywords but not fully processed by computer programs. Such automation would be beneficial for finding and matching offers and even for making transactions by

Przemysław Więch
Warsaw University of Technology, Institute of Computer Science
e-mail: P.Wiech@ii.pw.edu.pl

Tomasz Gidlewski
Warsaw University of Technology, Institute of Control and Computation Engineering
e-mail: T.Gidlewski@stud.elka.pw.edu.pl

M. Kaleta & T. Traczyk (Eds.): Modeling Multi-commodity Trade, AISC 121, pp. 99–111.
springerlink.com © Springer-Verlag Berlin Heidelberg 2012

autonomous agents. For this goal to be obtained, a knowledge management system is necessary for describing distributed and heterogeneous product and offer descriptions.

In the Semantic Web, ontologies provide a knowledge representation formalism to express the knowledge in the Web. Their role is also to enable different systems to communicate using a vocabulary with common semantics and to simplify reuse of gathered knowledge. Ontologies describe concepts, which are used to describe world objects, and relationships between concepts or objects. It is required that ontology languages have a well defined syntax and semantics, an adequate degree of expressivity and support for reasoning mechanisms.

Figure 1 shows the Semantic Web Stack as proposed by Tim Berners-Lee. Ontologies are a key element for knowledge representation and can be augmented by a query language such as SPARQL [10] and a rules language such as SWRL [7].

The Multicommodity Market Model M^3 is a well defined model for electronic trade. It contains XML schemas for describing market entities, commodities and complex offers. However, the current trends in the Semantic Web community focus on expressing all accessible knowledge in a common standard language. Today, the standard language for knowledge representation in the Semantic Web is OWL (Web Ontology Language). Being able to describe the M^3 model in OWL can bring many benefits resulting from the interoperability between the M^3 model and other resources in the Web.

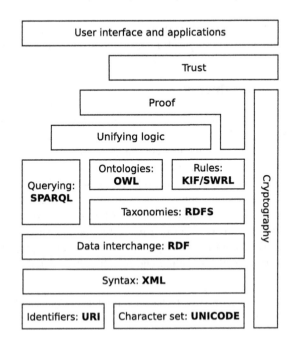

Fig. 1 Semantic Web stack

2   Ontologies

Ontologies provide a means for the integration of distributed and heterogeneous information sources. They are intended to provide a common vocabulary for computer systems to express information using the same concepts having the same semantics. Ontologies are used to describe objects and relationships between them. It is worth noting that ontologies are more than just vocabularies, dictionaries or hierarchies of concepts. They can express complex descriptions of real world objects and relationships between them. Unlike the practices of object-oriented programming, multiple inheritance is very common.

Today, the most often used ontology language is OWL (Web Ontology Language) [9]. It has a relatively high expressivity and a well defined semantics based on description logics. There is a well established base of reasoning algorithms for description logic-based languages [1].

3   Web Ontology Language

Ontology languages are created to build machine-readable descriptions of objects, classes and relations. There are several requirements for an ontology language. To be able to provide automatic reasoning services, the language has to have a formalized semantics, which is based on a logic formalism. It has to conform to open standards and be itself an open standard so that everyone can use and extend the format of descriptions. Moreover, a logical formalism has to be chosen to be able to express complex object descriptions and at the same time to manage to perform reasoning tasks in an acceptable time. An ontology language should provide the means to reuse previously created ontologies.

The current standard for the ontology language for the Semantic Web is OWL (Web Ontology Language) [9]. It is a W3C standard which is built on the experience of older languages like KL-ONE [11] and DAML+OIL [3]. The Web Ontology Language extends RDF [8] and XML [2] adding features for expressing formalized descriptions. There are three increasingly expressive sublanguages of OWL – *OWL Lite* providing simple features, *OWL DL* which has increased expressiveness without losing computational completeness and decidability, and *OWL Full* with the largest expressive power but with no computational guarantees. The most interesting from the point of view of e-commerce is OWL DL, which derives its name from the underlying formalism of description logics.

Description logics (DLs) [1] are a family of knowledge representation formalisms. Knowledge in DLs is represented by defining concepts from a selected domain, which comprise a terminology, and using these concepts for classifying objects and describing their properties. Description logics differ from earlier knowledge representation methods, such as semantic networks because they have a formalised semantics based on first order logic. This property of DLs enables reasoning using logic-based methods.

A knowledge base expressed in description logic consists of two components. The *terminology* (Tbox) is a set of axioms defining and describing *concepts* and *roles*. Complex concepts and roles are defined using basic ones. The *assertional component* (Abox) contains facts about objects expressed using concepts and roles.

4 Describing Offers with Ontologies

Ontologies are an important aspect in describing offers for products and services in e-commerce [4]. The business environment is inherently distributed and heterogeneous. Different entities on the market will want to see product and services data using their own perspective. An offer description consists of two parts: representation of the product or service and the details of the actual offer. These two aspects are interconnected but can nonetheless be developed separately.

Product and service categorization standards such as UNSPSC, eCl@ss and RosettaNet Technical Dictionary already exist. However, they do not comply with the ontology philosophy and cannot be treated as ontologies. However, they can be adapted to work with the Semantic Web infrastructure [5].

The most recent ontologies that deal with products, services stem from the already developed dictionaries. Examples of such ontologies are eClassOWL [5], which will be further described in the following subsection, and unspscOWL, which is has not been yet published.

When dealing with offer descriptions, apart from the product or service data we also need to specify the details of the actual offer, such as price, terms of delivery, etc. GoodRelations [6] is an example of an ontology that serves this goal. It can be used with external product and services ontologies and provides concepts necessary to construct an offer. The M^3 model also covers the issue of describing offers in terms of market entities, offer requirements and restrictions.

The offer ontology has to be prepared for different usage scenarios such as the following:

- A company offers to sell or rent a kind of product without identifying individual items.
- A person offers to sell a concrete instance of an item (e.g. used car).
- A Web resource (usually within the producer's Web site) describes the specifications of a product.
- Someone offers certain types of services for a described range of products (e.g. washing machine repair).

The following subsections describe ontologies, which can be used for extending the M^3 ontology to use publicly available reusable resources.

4.1   eClassOWL

The eClassOWL ontology is derived from the eCl@ss categorization standard [12]. The ontology provides a hierarchy of products and services concepts (e.g. *TV Set*),

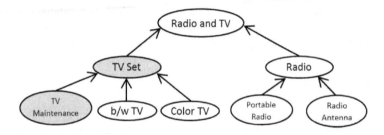

Fig. 2 eClassOWL hierarchy example.

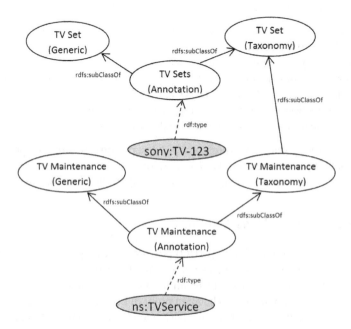

Fig. 3 eClassOWL ontology example.

product properties (e.g. *screen size*) together with enumerated values and recom-
mendations about the usage of properties and values.

The ontology is very large as it contains more than 25 thousand categories. Figure
2 shows an excerpt from the hierarchy. In this example it can be clearly seen that
the hierarchy is not designed to be interpreted as an *is-a* inheritance hierarchy, as
here *TV Maintenance* is under *TV Set* and *Radio Antenna* is under *Radio*. This tree
structure forms categories of intermixed products and services. As such it can not
be directly transformed into a valid OWL ontology.

Because of the structure of the original eCl@ss categorization, each category is
represented by three OWL classes. The *generic concept* should be treated as the
class of actual instances of products. The *taxonomic concept* corresponds to the

eCl@ss category hierarchy. The *annotation concept*, which is a subclass of both previous classes, is a convenience class to be used as the direct class of individuals. Figure 3 illustrates how the three types of concepts are linked together.

4.2   GoodRelations

Apart from annotating products and services, there is the need for describing details of individual offers. The GoodRelations ontology [6] fills this need by introducing concepts connected with the relationships between e-commerce entities, products, terms and conditions of contracts.

The primary concepts of the GoodRelations ontology are:

- Web Resource – a resource on the Web containing information about a business entity, offer, product, etc.,
- Business Entity – a legal body or person,
- Offering – an announcement of a *Business Entity* to provide a certain *Business Function* for a certain *Product or Service*,
- Business Function – the type of offer such as sale or rental,
- Product or Service – a product or service description.

Using this ontology one can express simple descriptions of offers of products or services made by a business entity. The ontology is not suitable for representing complex offers, such as multi-commodity ones, and relationships between them.

4.3   Measurement Units Ontology

The Measurement Units Ontology (MUO) provides a description of physical magnitudes by representing units of measurement in RDF. MUO consists of two parts. The first part describes how to define quantities with units, The second part is a huge repository of measurement units. The main profit of keeping units in the ontology is that quantities described by units from MUO can be easily converted to other units, which is very important in comparing parameters to each other. For example, the values *1.5 km* and *300 m* defined by MUO's units can be compared because derived units are defined as base unit and prefix. The structure of the ontology is shown in Figure 4.

To show how to represent physical magnitudes using MUO let us focus on an example. Our task is to represent information about a connection cable. We know that it is named ACLine1 and it has a length of 500 meters. Figure 5 shows this information represented in the ontology. The schema shows three types of objects: rounded rectangles, ovals and squares that represent classes, individuals and data properties, respectively. Dashed lines show object and data properties and solid lines represent ontological relations. The first row of objects in the schema is defined by MUO, while the remaining elements have to be additionally defined. This is the method used for adding ontological information about measurement units to the M^3 ontology.

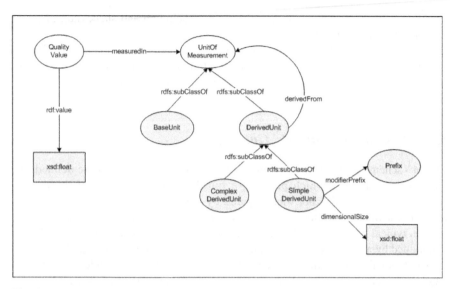

Fig. 4 The structure of MUO. Source: morfeo-project.org

5 Ontologies for M³

5.1 Motivation

The current version of the M³ model is implemented as an XML representation (M3-XML), which is defined by XML Schema definitions. The format has been introduced specifically for the needs of this model. The primary purpose of introducing ontologies into the M³ model is to improve the interoperability with other standards and tools.

The basic purpose of using an ontology in M³ is to introduce dictionaries for the M3-XML dialect. The feature of ontologies, which is valuable for M³ is that

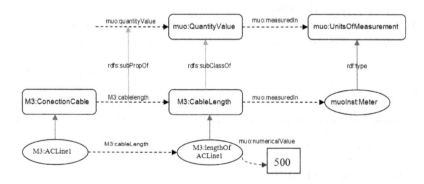

Fig. 5 Use of MUO in M3

specific relations can be easily described, e.g. concluding and succession of time periods, or relationship types between market entities and commodities. Further uses are related mainly to offer specification. Before sale negotiations, matching offers is very important, but when offers and commodities are described on different levels of generality, it can be difficult. Ontologies and reasoning engines can make the task easier.

5.2 Description of the M^3 Ontology

The M^3 scheme provides at least three ontology levels. The basics are defined by the M3Root ontology, which provides a full description language of the multi-commodity market. M3Root was prepared by the M^3 project. The next ontology level complements it by terms specific for each market dialect, rules, restrictions or even typical parameter and commodity definitions. This level should usually be maintained by a market operator or public department. The last ontology level contains individuals. Depending on the market type, this level can include one or more ontologies prepared by market operators, commercial brokers or market entities.

The M3Root ontology is based on the M^3 data model and has the same expressive power. The structure of the M3Root ontology is shown in Figures 6 and 7. The conversion mechanism had to take into account the differences between the OWL structure and the object data model.

5.3 Bidirectional Data Conversion between M3-XML and OWL

In order to introduce ontologies into the M^3 model, we propose to enable automatic conversion between the current M3-XML format into an ontology format, specifically into the OWL language. Since there is already some sample data accessible in M3-XML, one can take advantage of ontologies by automatically converting the available data into the new form. Although we see that the M^3 model could be fully expressed in OWL, we also see the need for it to remain backward compatible with M3-XML due to the fact that there are already implemented tools, which make use of the original format. This leads to the conclusion that an inverse conversion, from OWL to M3-XML, would also be beneficial. For the inverse conversion to be possible, the ontology representation has to preserve all information, which is expressible in M3-XML.

The converter has been fully implemented in Java 1.6 together with the following technologies. A DOM parser[1] was used to read M3-XML data. The interaction with the ontology was implemented with Jena Semantic Web Framework[2]. Pellet[3] was used to build inference models for the purpose of querying the ontology. Building

---

[1] http://www.w3.org/DOM/
[2] http://jena.sourceforge.net/
[3] http://clarkparsia.com/pellet/

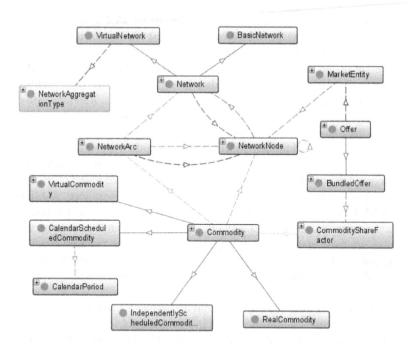

Fig. 6 Structure of M3Root ontology

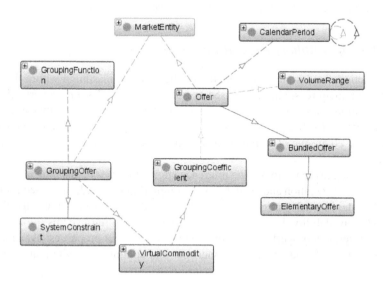

Fig. 7 Offers description in M3Root ontology

M3-XML files was based on Java classes generated by JAXB[4] from XSD[5] files. For viewing and editing OWL files Protégé[6] v. 4.1.0 was used with the following plug-ins: OntoGraf v. 1.0.1 and Pellet Reasoner v. 2.1.1.

The output of the conversion is in the form of a flat hierarchy, where all commodity kinds are transformed into direct disjoint subclasses of the Commodity class and individual objects are direct instances of these subclasses. References to other instances are mainly converted to object property assertions and attributes are represented in the ontology as data property assertions. In some cases a reference to another instance keeps additional information such as the type of relation between Market Entities and Network Nodes. This problem was overcome by defining each relation type in the ontology as a subproperty of the relatedTo object property. This construction makes it possible to state how the two objects are related. A difference between XML and OWL structures can be noticed in network definitions, where nodes and arcs are placed inside a network description. A network membership relation does not exist in the ontology language, so it has to be replaced by an object property joining the network and it's parts. The definitions of generic parameters and particular values are described using the MUO ontology, which was presented Section 4.3.

It is important that the data imported into an ontology must be consistent. This leads to the conclusion that inserting data into ontology is sensitive to the order of files and nodes processing. This happens because to build a relation between a pair of objects, their representation must already exist in the ontology. This means that converting a collection of NetworkArcs before NetworkNodes cannot be performed, as a NetworkArc cannot exist without its start and end. Problem mentioned above leads to need of consistency checking in the conversion process.

5.4   Use of Ontology Reasoning for M^3

In this section we describe the potential applications of reasoning within the M^3 project. The main reasoning tools used in following examples are property chains and anonymous class membership. A property chain is a way of composing several properties of subsequent objects into one named property. If we define r3 as the property chain r1 ∘ r2, it means that from r1(A, B) and r2(B, C) we can infer r3(A, C).

The first suggestion is to use the inference engine in M^3 for reasoning about virtual network aggregation and availability of commodities. The idea is to deduce real network nodes in witch particular commodity is available from the knowledge about commodity availability described in a virtual network. To retrieve this information we should define a property chain consisting of two object properties: availableAt and aggregates − Node, and the reasoner will join commodities with real network nodes.

---

[4] https://jaxb.dev.java.net/

[5] http://www.w3.org/TR/xmlschema-1/

[6] http://protege.stanford.edu/

In the electrical energy market, a network consists of energy producers, receivers and connections between them. Producers offer commodities located in their own nodes and market operator set up a Flowgate Right (FGR) on network arcs. For clients, it is important where each operator offers his energy. Such an operation can be obtained by using two property chains. The first one joins electric energy commodities with network arcs with FGR related to the producer's origin node. Another property chain takes the information found about ends of recently discovered arcs, and joins them with particular commodities.

Another example deals with finding offers based on time intervals. This requires changing the representation of time periods in the ontology. Currently, periods are described as data properties specifying their start and end. To reason about intersections of periods, it is needed to keep every simple time unit as an individual. Every period must have a defined predecessor and successor. Basing on this information, the reasoner creates intersections between all given periods. One of the disadvantages of such a description is the large size of the data, with which reasoners can have problems. There are two ways of finding overlapping periods. The first method involves choosing disjoint periods (by one simple property chain) first and then complementing the obtained set. The other way requires using two property chains for comparing the start and end time instances and resulting in these periods that satisfy conditions of both property chains.

6   Problems with Ontologies

There are several issues that might be problematic when using ontologies for offer descriptions. One of the problems is the use of the existential quantifier. In order to express the fact that *there exists an $\mathcal{X}$ such that $\mathcal{X}$ is a 40-inch TV set and company* ABC *sells $\mathcal{X}$* one has to use the existential quantifier which introduces an anonymous instance. Such constructs have a substantial negative impact on computations speeds of reasoning engines. One of the methods to work around this problem is to use special instances which convey the meaning of *object type*.

The OWL ontology language forces using unique URI identifiers for all classes, properties and objects, which will be referred to, e.g. `http://www.example.com/products#ExampleProductType`. It is not always convenient to reserve a domain name, which should physically exist, to create ontologies, product and offer descriptions.

The heterogeneity of ontologies in the Web with different entities using different vocabularies, and building ontologies from their own point of view leads to a necessity to integrate heterogeneous information sources. Two major solutions are sought to deal with this issue. Firstly, standardization of upper ontologies dealing with abstract concepts common for different ontologies is a necessary step for agents to function in the distributed environment of the Web. However, the heterogeneity of ontologies is inevitable as, for example, various manufacturers may want to name and describe their technologies in their own terms for marketing purposes. In this case standardization might help but will not solve the problem. The second solution here is to integrate the ontologies by matching their concepts using a relation such as equivalence or subsumption.

7 Conclusion

The Semantic Web is emerging as a web of semantically described and interconnected pieces of knowledge expressed in a standard representation language. Ontologies, which are the main knowledge representation formalisms, have the potential to change the way offers and concepts related to them are expressed in the Internet. Using a formal representation language has the benefit of automated reasoning. This in turn can provide means for tasks such as finding and matching offers or making transactions by autonomous agents.

We have described several ontologies, which can be treated as complementary for describing offers. The eClassOWL provides a categorization of products and services and the GoodRelations ontology overlaps with the M^3 model and provides descriptions of market entities and offers. The Measurement Units Ontology provides a good way of describing values together with the unit of measurement. We have shown how to use this ontology in the M^3 model.

We have proposed a translation of the M^3 model and the M3-XML data format into an ontology expressed in OWL. Some of the conversions are not straightforward mappings from the XML format, however the OWL representation does not introduce any data loss. We have also shown exemplary reasoning tasks, which could be accomplished with standard description logic reasoners.

References

1. Baader, F., Calvanese, D., McGuinness, D.L., Nardi, D., Patel-Schneider, P.F. (eds.): The description logic handbook: theory, implementation, and applications. Cambridge University Press, New York (2003)
2. Beckett, D.: RDF/XML syntax specification (revised) (2004),
   http://www.w3.org/TR/rdf-syntax-grammar/
3. Connolly, D., van Harmelen, F., Horrocks, I., McGuinness, D.L., Patel-Schneider, P.F., Stein, L.A.: DAML+OIL reference description (March 2001),
   http://www.w3.org/TR/daml+oil-reference
4. Fensel, D., McGuinness, D.L., Schulten, E., Ng, W.K., Lim, E.P., Yan, G.: Ontologies and electronic commerce. IEEE Intelligent Systems 16(1), 8–14 (2001)
5. Hepp, M.: Products and services ontologies: A methodology for deriving OWL ontologies from industrial categorization standards. Int. J. Semantic Web Inf. Syst. 2(1), 72–99 (2006)
6. Hepp, M.: Goodrelations: An ontology for describing products and services offers on the web. In: EKAW, pp. 329–346 (2008)
7. Horrocks, I., Patel-Schneider, P.F., Boley, H., Tabet, S., Grosof, B., Dean, M.: SWRL: A semantic web rule language combining owl and ruleml. Tech. rep., World Wide Web Consortium (2004)
8. Manola, F., Miller, E.: RDF primer (2004),
   http://www.w3.org/TR/rdf-primer/

9. McGuinness, D.L., van Harmelen, F.: OWL web ontology language overview (2004),
   http://www.w3.org/TR/owl-features/
10. Prud'hommeaux, E., Seaborne, A.: SPARQL query language for rdf (2008),
    http://www.w3.org/TR/rdf-sparql-query/
11. Schmolze, J.G., Beranek, B., Inc, N.: An overview of the KL-ONE knowledge represen-
    tation system. Cognitive Science 9, 171–216 (1985)
12. International standard for the classification and description of products and services,
    http://www.eclass-online.com/

Reliability Aspects of Multi-commodity Markets

Jacek Malinowski

Abstract. In this paper several reliability aspects of multi-commodity trade are discussed. As it is very often required that trading decisions should be taken both rationally and very quickly, e.g on a short-term electric power market, in such cases trading is performed automatically by multi-agent systems. Thus the question of MAS reliability arises, which is herein considered in two aspects – topological and functional. Interactions of agents in a MAS occur according to a certain topological pattern which can be directly transformed to the structural reliability model of that MAS. A number of such topological patterns are presented along with respective reliability models. In its functional aspect a MAS can be seen as a graph whose nodes process input information and pass it to other nodes thus fulfilling collectively certain task. An agent's malfunction can lead to a delay, misfulfillment or failure of that task. An interesting model of inter-agent functional dependence based on game-theoretical approach is presented. Apart from reliable execution of trading operations, an important issue is the ability to quickly assess whether the fulfillment of a contract is technically possible. An example is given where the effective conditions which energy suppliers' capabilities must meet are defined as the constraints of a transportation problem. Also, a non-trivial problem of quickly finding those constraints is addressed.

1 Introduction

In general sense reliability is a feature that can be described as proper and faultless functioning of a system composed of multiple interacting components for the purpose of fulfilling specific tasks. Reliability can be considered either in quantitative or qualitative sense, as shown in Fig. 1 presenting a rough classification of various approaches to the concept of reliability.

Jacek Malinowski
Systems Research Institute, Polish Academy of Sciences
e-mail: Jacek.Malinowski@ibspan.waw.pl

M. Kaleta & T. Traczyk (Eds.): Modeling Multi-commodity Trade, AISC 121, pp. 113–125.
springerlink.com © Springer-Verlag Berlin Heidelberg 2012

The above definition can also be applied to multi-commodity market. However, with regard to various aspects of multi-commodity trade, reliability acquires a number of specific meanings. The aspects in question are: the object of trade, the assumed model of trade, procedures and technical means used for negotiating, settling and fulfilling a contract, or economical, political, and natural circumstances. Each of them carries a risk factor which can be tentatively defined as the probability (possibility) of non-fulfillment of intended purpose (e.g. failure to fulfill a settled contract due to a contractor's or an agent's fault) or the occurrence of undesired event (e.g. an outage in a power distribution network, if electric power is the object of trade).

Probabilities (possibilities) related to individual risk factors can be regarded as basic reliability indices of a system understood as the set of subjects (contractors and/or agents) participating in the purchase, sale or exchange of commodities or services conducted according to specific procedures, with the use of specific technical means, in given external circumstances. Basic indices, apart from providing information, which allows for basic assessment of various types of hazards, also constitute a starting point for determining secondary indices derived from primary ones. like the parameters of a stochastic process describing the course of events which occur during trade proceedings, distribution function of the time to the occurrence of undesired event, or the availability of one or some of the system's components. As far as multi-commodity trade is concerned, the main function of reliability indices is the assessment of risk related to multiple factors occurring in the course of commercial dealings. Such assessment, correctly performed, is virtually necessary for taking proper decisions at consecutive stages of preparing and fulfilling contracts. Besides, on its basis one can forecast the cost-effectiveness of a contract, or to a certain degree safeguard oneself against possible losses, e.g. by buying an insurance policy.

The main goal of market mechanisms is to secure the demand-supply equilibrium (see [3]), therefore balancing these two factors constitutes a fundamental market process. The investigation of this process on account of various types of irregularities (e.g. lack of balance for a specific commodity, group of commodities, or on a given territory, faulty performance of agents) is another essential element in reliability analysis of market systems.

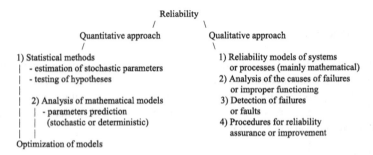

Fig. 1 Concept of reliability – classification

Determination of reliability characteristics must be preceded by accurate, multi-aspect analysis of a given trading system, both qualitative and quantitative. For this purpose a number of standard methodologies and tools can be applied, e.g. FMEA/FMECA, HAZOP, decision trees or diagrams for qualitative analysis, and a plethora of mathematical tools for quantitative analysis: reliability theory, probability theory and statistics, stochastic processes theory, fuzzy sets theory, fault trees, reliability block diagrams, Petri nets, Bayesian networks. However it should be taken into account that each system has its own, often unique, specific properties which render it necessary to adapt standard methods, or develop and implement specialized ones.

The notion of reliability is akin to several other concepts, namely dependability – reliance on proper functioning, survivability – ability to operate in spite of incurred damages, availability – probability of proper functioning at a given moment or the mean value of this probability on a certain time interval, fault tolerance – continuity of correct operation regardless of components failures, durability – ability to operate for sufficiently long time.

2   Multi-Agent Model of Multi-commodity Trade

A multi-agent system (MAS) is a system composed of multiple intelligent entities (agents) operating interactively for the purpose of fulfilling specific tasks which cannot be realized by a monolithic (single module) system. An agent is an autonomous component of the system, it operates in certain environment, monitors this environment, and reacts to events occurring in it. Additionally, it can adapt itself to the changing external conditions, which means that it can change the way it responds to the occurring events. This definition (a modification of the one given in [4]) originates from computer science, but it is so general that it can be applied to various types of systems: computer, manufacturing, economic, public administration, socio-political, as well as stocks, commodities or consumer goods trading systems. For example the authors of [3], constructing a model of local electric power market, distinguished five types of agents: balancing operator (BO), trading agent (TA), physical agent (PA), external market operator (EMO), and local market operator (LMO). BO organizes and implements the central balancing process. TA and PA represent buyer or seller, the former models the activity of a market participant, the latter – the functioning of equipment and the occurrences of natural phenomena in their immediate proximity. EMO simulates the impact of external (global) markets on the local market under consideration. LMO organizes the functioning of the local market, inter alia, it manages the agents register and provides information regarding natural phenomena. Figure 2 (taken from [3]) illustrates the flow of information and the interactions between the described system's agents. For broader introduction to multi-agent systems the reader is referred to [5].

Although this work is focused on multi-commodity markets, the question of reliability is considered herein in broader sense, i.e. in relation to generic multi-agent systems. Thus, the multi-agent market model should be viewed as a special case of

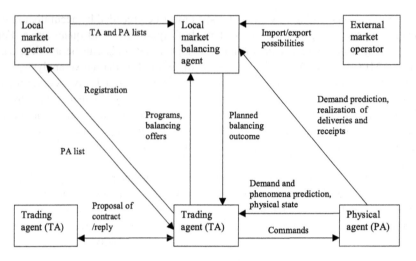

Fig. 2 Schema of agents cooperation in an exemplary MAS

a generic multi-agent system. The MAS reliability will be investigated in two basic aspects – functional dependence between MAS components, and the fulfillment by MAS of the assigned task.

Remark: In the sequel we will distinguish between a multi-agent system (referred to as MAS), and a system in the reliability sense – a set of components whose performance determines the system's operational state.

3   Topologies of Interaction in Multi-Agent Systems

The interaction of agents in a MAS takes place according to certain schemas that can be illustrated on diagrams whose topologies render typical policies according to which the agents interoperate. We can distinguish four basic topologies – web-like, star-like, grid-like, and hierarchical-collective agent network. On the following figures the topologies of agents' interaction are depicted as graphs, agents being represented by nodes, and interaction relations – by links.

The web-like topology is used to model environments in which every agent can directly interact with any other agent in the system. It is represented as a complete graph, see the example in Fig. 3[1].

In a system with web-like topology there is no hierarchy among agents, they are identical as to the internal structure, knowledge of the environment, and the set of actions. The choice of appropriate action is based on the same decision process. However, the agents can differ as to the sensory inputs and currently taken decisions, because they can find themselves in different environmental conditions. Admittedly, they have equal capabilities and operate according to one decision procedure, but

---

[1] All figures in this section are based on [6].

they only have limited (incomplete) information about other agents' internal states and sensory inputs. Thus, they are not able to predict other agents' actions.

The reliability model adequate to web-like topology is the k-out-of-n:G system, also known as voting system. Such system is composed of n independent components (the independence means that the state of a component does not affect the states of the other components), and for its proper operation it requires at least k components to be functional. If the number of failed components exceeds $n - k + 1$, then the system is inoperable. However, the adequacy holds under the following conditions: the MAS in question includes only one group of mutually communicating agents, and performs its tasks properly if the sufficient number of agents are available and fully functional. The weakening of these assumptions may result in greater complexity of the reliability model which is likely to become an extension of the k-out-of-n:G system

In a MAS with the star-like topology the actions of agents are supervised and/or coordinated by supervising agents named supervisors, coordinators or facilitators. Agents can cooperate with other agents only if they communicate supervisors first. A supervisor acts as intermediary in activating agents and sending data between them. There can be multiple coordinators in a MAS, the exemplary star-like topology is shown in Fig. 4.

In frequently encountered situations agents are divided into groups, each of which executes different tasks. In such configuration agents do not communicate directly between different groups, or even within a group, but by means of the intermediary agent which assigns tasks to individual agents and coordinates task execution. The star-like topology is best suited to describe relations between agents in such systems.

While designing a MAS it is essential to decide whether agents are benevolent (cooperative) or competitive. Even if agents have different goals, the benevolent

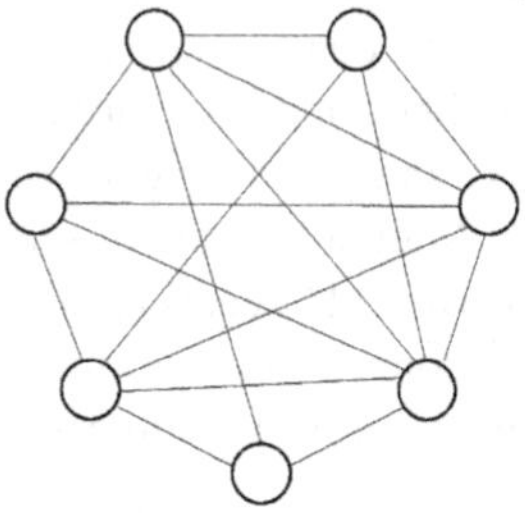

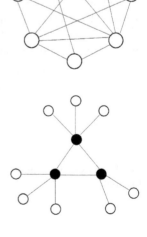

Fig. 3 An example of web-like topology

Fig. 4 An example of star-like topology

ones can help other agents attain their goals, especially if they are not in conflict with their own. Otherwise agents are "selfish" which means that they attempt to reach only their own goals. In the extreme case agents can play a "zero-sum" game, i.e. deliberately act to the detriment of the other agents. The star-like topology seems most adequate for performing arbitration in such conflict situations.

When constructing a reliability model of a MAS with this type of topology one must take into account the existence of "single points of failure" which the supervising agents undoubtedly are. Hence, the first approximation of the reliability structure of such a MAS is the series system composed of modules which correspond to separate groups of agents. Obviously, each such group has its internal structure which, when taken into account, may significantly increase the model's complexity. On the other hand, enhancing the system's reliability by introducing redundant supervising agents does not complicate the model significantly – a series system becomes a series-parallel one.

In a grid-like MAS each agent directly interoperates only with neigboring agents – the neighborhood is understood in the functional rather than geographical sense. A group of neigboring agents is called Agency. It can be managed by a coordinating agent assigned for this purpose. Interactions with agents from outside of the neighborhood are possible only through the coordinator – if one is present. Such agent needs not directly communicate with all other coordinators in a MAS, hence an interaction between two coordinators may take place with the participation of a number of other ones. The role of coordinator can be "dynamically" assigned or revoked in a way ensuring maximal efficiency of interaction between agents at a given moment. An exemplary grid-like topology is shown in Fig. 5. Agents denoted by filled circles play (temporarily) the role of coordinators.

The reliable performance of a grid-like MAS first of all requires that each coordinator function properly and each two of them can communicate (possibly indirectly). The basic reliability model of a MAS of this type is the graph representing the configuration of (functional) connections between coordinators. Such model will become even more complex after incorporating the dynamics of the process of assigning/revoking the role of coordinator to agents.

Main attributes of the fourth basic interaction topology – HCAN – are the following: (1) agents are grouped in layers, (2) layers form a hierarchical structure, (3) no connections (in functional sense) exist between agents within the same layer, (4) agents from neighboring layers communicate according to peer-to-peer paradigm,

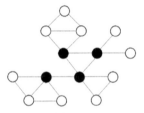

Fig. 5 An example of grid-like topology

(5) interactions between agents in a given layer are coordinated by agents from the immediately higher layer. An example of HCAN topology is presented in Fig. 6.

The reliability model well suited to HCAN topology is apparently a graph with the structure that can be described as "interlaced trees", i.e. having common nodes and links. Obviously, it should be assumed that each common node of those trees is not a root node of any of them.

It is nearly certain that the topologies presented in the current section do not exhaust all possible configurations of agents interactions. However, if it occurs that the functioning of a given MAS cannot be described using one of those topologies, then it should be advisable to attempt a combination of two or more basic ones.

The common characteristic of herein presented reliability models is their structural aspect. Clearly, this is a consequence of the topological description of MAS functionality. Using such models one can determine multiple "static" reliability parameters. The term static means that they refer to a fixed point in time, hence they do not reflect the dynamical properties of a considered system. Obviously, assuming that the topology can change in time (as in the grid-like case) paves way to determining dynamic (time-dependent) parameters. As far as qualitative approach to reliability is concerned, the considered models can serve as a starting point for designing dependable MAS or enhancing their performance.

4 Game Theoretical Approach to MAS Functionality

And interesting and promising approach to the research on MAS reliability is the analysis of MAS functionality by means of game theory. More accurately, a so called game of deterrence is used to model the behavior of a MAS (the deterrence relation between strategies of a game is defined later in this section). One of the primary sources on this matter is [2]. The functioning of a MAS is therein presented as a system of propositions (in the sense of propositional calculus) put together in a graph structure which defines certain game of deterrence. This game's playability indices provide information about the agents reliability, and in case of an agent's failure can be helpful in the search for the source of failure. In the cited source the question of MAS reliability enhancement by means of adding redundant agents is also considered.

After this theoretical introduction a simple explanatory example is in place here. Let the system be composed of three agents: p, q, and r. The agent p

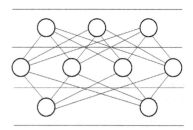

Fig. 6 An example of HCAN topology

processes certain input data, transfers it to the agent q, which (after further processing) transfers it to the agent r. It is assumed that all agents are subject to internal failures. Let P, Q, R, Iq, and Ir be propositional variables defined as follows:

P : p is operable
Q : q is operable
R : r is operable
Iq : internal failure of q has occured
Ir : internal failure of r has occured

The graf of propositions, which represents the functioning of the above MAS is shown in Fig. 7[2].

The formula "A > B" denotes the implication "if A then ¬ B", which means that "A excludes B" (¬ is the negation operator). Notice that > is not the implication operator. "Iq <> ¬ Iq" is the so-called consistency condition saying that the events Ix and ¬ Ix cannot occur simultaneously. This condition is introduced so that Iq and Ir will not be root nodes of the above graph. If they were this would mean that the internal failures of q and r always happen. Obviously, it is necessary to admit the possibility that this does not happen, but also the possibility that it does. The lack of the consistency condition for P has two reasons. First, our aim is to assess the influence which the proper (not improper) functioning of p exerts on the other agents. Second, such condition would be partly redundant to P, as P has no predecessor.

It has been proved in [2] that each system of propositions presented in the form of a graph can be mapped to exactly one two-player matrix game of deterrence. The strategies of player 1 correspond to "odd" vertices of the graph and consist in generating the events like "agent functions properly" or "agent's internal failure has not occurred", while the strategies of player 2 correspond to "even" vertices and consist in generating the events like "agent does not function properly" or "agent's internal failure has occurred". Thus each two adjacent vertices of the graph correspond to different players' strategies. This interpretation applied to the graph in Fig. 7 yields a game of deterrence with the following sets of strategies:

for player 1 : {P, Q, ¬ Iq, ¬ Ir, R}
for player 2 : {¬ P, Iq, ¬ Q, Ir, ¬ R}

For strategies in (non-fuzzy) matrix games one defines the so-called positive playability index which is equal to one if (1) the strategy guarantees positive payoff

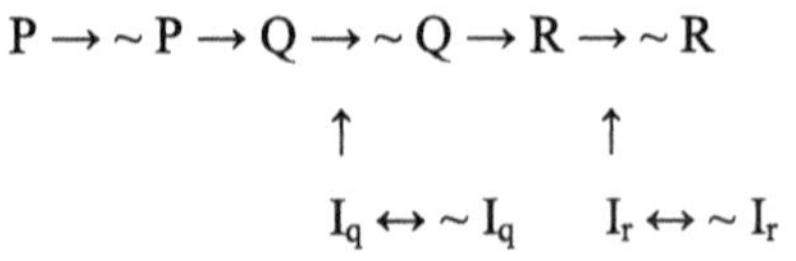

$$P \rightarrow \sim P \rightarrow Q \rightarrow \sim Q \rightarrow R \rightarrow \sim R$$

$$\uparrow \qquad\qquad \uparrow$$

$$I_q \leftrightarrow \sim I_q \qquad I_r \leftrightarrow \sim I_r$$

Fig. 7 The graph for the above defined MAS

_____

[2] All figures in this section are based on [2].

irrespectively of the oponent's strategy, or (2) the strategy guarantees positive payoff provided that the oponent chooses a strategy with positive payoff or, if such strategy does not exist, any strategy. Otherwise that index is equal to zero. Clearly, for a non-fuzzy game positive playability indices are binary numbers from the set $\{0, 1\}$. A game is called fuzzy if its positive playability indices can take values from the interval $[0, 1]$. Hereafter the positive playability index of the strategy s will be denoted as $J(s)$. In case of the fuzzy game describing the functioning of the considered three-agent system strategies are identified with specific events, hence if X denotes the event "the agent x is in operable state" then $J(X)$ is a measure of possibility of the occurrence of X, provided that the agent p is in operable state.

The graph in Fig. 7 depicts the so-called deterrence relations occurring between some strategies. In general case, in a two-player matrix game a strategy x followed by player 1 deters player 2 from choosing strategy y, if

1) $J(x)>0$, or $J(a)=0$ for every strategy a from the strategies set of player 1,
2) if player 1 follows x then choosing y is not profitable for player 2,
3) in the strategies set of player 2 there is a strategy b such that $J(b)>0$.

In game theoretical description of MAS functionality the exclusion relation between events becomes the deterrence relation between strategies, thus, for example, "$\neg P > Q$" should be interpreted as "following the strategy $\neg P$ by player 2 deters player 1 from choosing the strategy Q".

For every two-player matrix game the positive playability indices of both players' strategies fulfill certain system of equations. The rules according to which such system is constructed for a fuzzy two-player matrix game with deterrence are given in [2]. For the game corresponding to the graph in Fig. 7 such system is composed of the following equations:

$$J(P) = 1 \; ; \; J(\neg P) = 0$$
$$J(Q) = [1 - J(\neg P)][1 - J(Iq)]v \; ; \; J(\neg Q) = [1 - J(Q)]v$$
$$J(R) = [1 - J(\neg Q)][1 - J(Ir)]v \; ; \; J(\neg R) = [1 - J(R)]v$$
$$J(Iq) = [1 - J(\neg Iq)]v \; ; \; J(\neg Iq) = [1 - J(Iq)]v$$
$$J(Ir) = [1 - J(\neg Ir)]v \; ; \; J(\neg Ir) = [1 - J(Ir)]v$$
$$1 - v = [1 - J(\neg P)][1 - J(\neg Q)][1 - J(\neg R)][1 - J(Iq)][1 - J(Ir)]$$

where $1 - v$ is the so-called default playability index for player 2. This index is equal to one, if the positive playability index is equal to zero for every strategy of player 2. This means that none of player 2's strategies is profitable, so that it makes no difference to player 2 which strategy to choose. In [2] it was proved that

$$J(Iq) = J(\neg Iq) = J(Ir) = J(\neg Ir) = v/(1+v)$$

which means that the chances of internal failure are equal for the agents q and r, so that

$$J(Q) = J(\neg Q) = v/(1+v)$$
$$J(Q) = J(\neg Q) = v/(1+v)$$
$$J(R) = v/(1+v)^2 \; ; \; J(\neg R) = v[1 - [v/(1+v)^2]]$$

and

$$1 - v = (1 + v - v^3)/(1+v)^5$$

The above system of equations has one non-binary solution, namely $v = 0.973$. In consequence, we obtain:

$$J(P) = 1 \; ; \; J(\neg P) = 0$$
$$J(Iq) = J(\neg Iq) = J(Q) = J(\neg Q) = J(Ir) = J(\neg Ir) = 0.493$$
$$J(R) = 0.25 \; ; \; J(\neg R) = 0.73$$

Now let us find, how the presence of an agent x redundant to q will increase the value of the positive playability of R. To do this, we have to analyze the graph in Fig. 8.

It occurs that the positive playability indices for P, ¬P, Q, Iq, ¬Iq, Ir, and ¬Ir are the same as in the system without redundancy. Furthermore we have:

$$J(X) = J(Q) = v/(1+v)$$
$$J(\neg X \wedge \neg Q) = [1 - J(X)][1 - J(Q)]v = v/(1+v)^2$$
$$J(R) = [1 - J(\neg X \wedge \neg Q)][1 - J(Ir)]v = [1 - v/(1+v)^2]v/(1+v)$$

As a result, $J(R) = 0.37$, $J(\neg R) = 0.612$. Thus placing the agent x in paralel with the agent q increases the value of J(R) by nearly 50%.

It seems purposeful to attempt applying herein presented method of MAS reliability analysis to more complex systems. It may however occur that the systems of equations for positive playability indices are so extensive that solving them "manually" is out of the question. Constructing an algorithm which could solve such equations automatically may be a way of overcoming this difficulty. It is also worthwhile investigating whether limiting considerations to only certain types of MAS structures may result in relatively easily solvable equations.

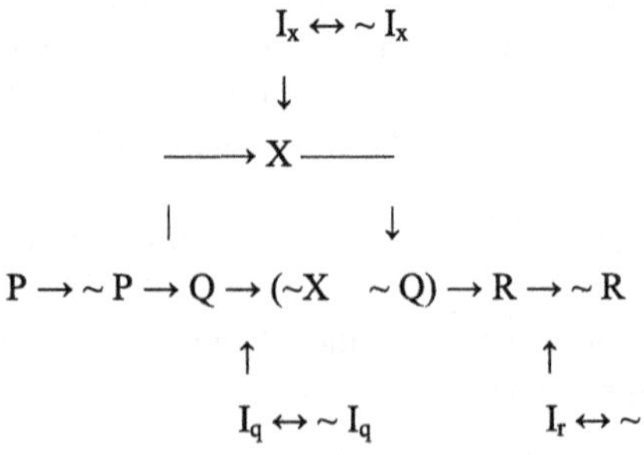

Fig. 8 Agent redundant to q added to the considered MAS

It should be noted that, considering the qualitative approach to MAS reliability, the methodology presented in this section can also be applied to the analysis of the causes of MAS improper functioning, or to the development of procedures ensuring MAS reliability improvement.

5   Analysis of Technical Conditions for the Fulfillment of a Contract

When negotiating a contract it is necessary to take into account the technical possibility, or lack thereof, of the fulfillment of the contract's terms. It can turn out that the capabilities of a commodity producers, and/or the technical infrastructure used in the conveyance of that commodity from producers to recipients, impose certain limitations enforcing particular ways of action in all stages of contractual procedures. Such limitations often have the form of linear inequalities defining necessary and sufficient conditions which must be fulfilled by producers' capabilities to meet the requirements of individual recipients. In general case constructing such a set of inequalities is a non-trivial task which can be formulated as follows: on the basis of primary (necessary) conditions stating which producers supply given commodity to particular recipients, create the list of effective (necessary and sufficient) conditions which the producers' capabilities must fulfill in order to satisfy the needs of all recipients. Clearly, the above task is related to the well-known transporation problem – the problem's constraints are non-conflicting if and only if all effective conditions are fulfilled.

For illustration let us consider the following example. It is assumed that three electric power producers, with the production capabilities W1, W2 and W3 respectively, supply the energy to four recipients with demands Z1, Z2, Z3, Z4. The scheme of power network connecting producers to recipients can be found in Fig. 9.

The network's structure determines that recipient 1 is supplied by producers 1 and 2, recipient 2 – by suppliers 1 i 3, recipient 3 – by suppliers 2 i 3, and recipient 4 – by supplier 3. This can be translated into the following primary conditions:

$$W1 + W2 > Z1$$
$$W1 + W3 > Z2$$
$$W2 + W3 > Z3$$
$$W3 > Z4$$

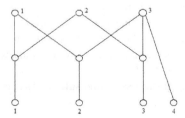

Fig. 9 The scheme of a simple power transmission network

Analyzing the network's structure and two first primary conditions we obtain the following dependencies:

$W1 + W2 > Z1$
$W1 + W3 > Z2$
$W1 + W2 + W3 > Z1 + Z2$

The third inequality follows from the fact that recipients 1 and 2, taken collectively, rely on suppliers 1, 2 and 3, hence the total energy produced by those suppliers must be greater than or equal to the total demand of recipients 1 and 2. Note that the obtained conditions are independent, i.e. none of them is the implication of one or more remaining ones. After taking into account the third primary condition, and applying analogous reasoning, the above list of conditions is transformed to the following one:

$W1 + W2 > Z1$
$W1 + W3 > Z2$
$W1 + W2 + W3 > Z1 + Z2 + Z3$
$W2 + W3 > Z3$

Finally, after the fourth basic condition is taken into account, the following list of five effective conditions is obtained:

$W1 + W2 > Z1$
$W1 + W3 > Z2 + Z4$
$W1 + W2 + W3 > Z1 + Z2 + Z3 + Z4$
$W2 + W3 > Z3 + Z4$
$W3 > Z4$

The transportation problem is well established in the optimization theory and has been extensively researched, however the construction of an efficient algorithm finding the list of effective conditions (constraints) in the context analogous to that presented above is still a task to be solved. It should be remarked that this is not a trivial problem.

Besides the limitations implied by the relations between producers and recipients one should also consider those which are imposed by the transmission/distribution infrastructure. In the given example such limitations are related to maximum capacities of power transmission lines. Thus, the capacity of the line ending at recipient 1 must meet demand Z1, total capacity of the lines which begin at producers 1 and 3, and run in the direction of recipient 2 must meet demand Z2, etc. This analysis results in the set of primary conditions, which, after applying the aforementioned algorithm used for determining minimum production capabilities, are transformed to the set of (necessary and sufficient) effective conditions that the lines' transmission capacities must fulfill.

It should also be taken into consideration that the components of transmission/distribution infrastructure are subject to random failures. The risks related to such failures should be taken into account in the process of contract settlement, but, usually, their assessment is not easy. It is therefore purposeful to perform a simulation of the failure-and-repair process of the infrastructure components, followed by the estimation of appropriate parameters of such process, thus obtaining data for the required assessment of risks.

References

1. Kwang, H.L.: First course on Fuzzy Theory and Applications. Advances in Soft Computing. Springer (2005)
2. Rudnianski, M., Bestougeff, H.: Multi-agent Systems Reliability, Fuzziness, and Deterrence. In: Hinchey, M.G., Rash, J.L., Truszkowski, W.F., Rouff, C.A. (eds.) FAABS 2004. LNCS (LNAI), vol. 3228, pp. 41–56. Springer, Heidelberg (2004)
3. Toczylowski, E., et al.: Electrical energy markets modelling. Selected issues. Systems Research Institute of the Polish Academy of Sciences (2007) (in Polish)
4. Toczylowski, E., et al.: Multiagent framework for trading on local electrical energy market. Design issues. Systems Research Institute of the Polish Academy of Sciences (2009)
5. Wooldridge, M.: An Introduction to MultiAgent Systems. John Wiley and Sons (2009)
6. Quiming, Z.: Topologies of agents interactions in knowledge intensive multi-agent systems for networked information services. Advanced Engineering Informatics 20, 31–45 (2006)

Part II
M³ Applications

M³ Applications on the Electricity Markets

Mariusz Kaleta, Przemysław Kacprzak, Kamil Smolira,
and Eugeniusz Toczyłowski

Abstract. The electrical energy markets illustrate very well complexity of the infrastructure markets. Achieving the market balance, while maintaining its high efficiency, is a challenging task which requires both a rich commodities structure and a complex structure of the market processes. Market balancing is being obtained by multi-step processes of the multi-commodity trade. In this chapter we show that the fulfillment of various specific electricity market requirements is possible by applying model M³. We focus on the balancing segment of the market and present relations and mappings between various concepts of a real balancing market and notions of M³. We discuss some potential ways of market development to demonstrate how M³ notions facilitate modeling of the new solutions.

1 Introduction

The power energy markets have many specific properties. The storage of the electrical energy is severely limited and involves high energy losses and therefore power demand and generation must be balanced in real time under severe security of supply requirements. Another issue involves physical limitations of the transmission

Mariusz Kaleta
Warsaw University of Technology, Institute of Control and Computation Engineering
e-mail: M.Kaleta@ia.pw.edu.pl

Przemysław Kacprzak
Warsaw University of Technology, Institute of Control and Computation Engineering
e-mail: P.Kacprzak@elka.pw.edu.pl

Kamil Smolira
Warsaw University of Technology, Institute of Control and Computation Engineering
e-mail: K.Smolira@elka.pw.edu.pl

Eugeniusz Toczyłowski
Warsaw University of Technology, Institute of Control and Computation Engineering
e-mail: E.Toczylowski@ia.pw.edu.pl

M. Kaleta & T. Traczyk (Eds.): Modeling Multi-commodity Trade, AISC 121, pp. 129–147.
springerlink.com   © Springer-Verlag Berlin Heidelberg 2012

network usage. For a typical commodity transferred within a network from a delivery point to a destination point, it is usually possible to choose a single path that commodity flows over the infrastructure network. The electrical power flows over the whole transmission network according to the physical (Kirchoff's) laws, and not just along single paths between delivery and destination points, as determined on markets by bilateral contacts. The System Operator (SO) has rather a limited control over the power generation and power flows in the transmission network. Finally, the demand side is still rather inelastic. All of these properties must be reflected in complex balancing processes in the electrical energy markets that require a rich commodities structure for defining various market products, not only for energy, but also for the transmission and other ancillary services, such as reserves.

The observed evolution of the energy market designs shows that current solutions to the balancing mechanisms must evolve, due to variability of the conditions in which power systems operate and increasing knowledge and experience. The market solutions must be adapted to current local conditions, including requirements and expectations of the market participants, which are naturally evolving. It seems unlikely that all the necessary changes for achieving a long-term, significant improvement in efficiency of the electricity markets could be made on the basis of "shock therapy". A more realistic approach involves evolutionary developments of both, the commodity composition and the market mechanisms. Many proposals for specific market solutions are formulated, and there is a plethora of various market designs. Each proposal usually contains many variants, and selection of the best specific variant is also challenging. It is therefore essential to create flexible conditions for evolutionary development of the market mechanisms.

In this chapter we present possibilities and advantages of using the multi-commodity approach with M^3 standard in electricity markets. Because of the complexity of the whole market design, we will focus just on the balancing market. We show that the multi-commodity market model allows to alleviate barriers for evolution through flexibility and openness of the design concept.

2   Market Processes in Electricity Market

Balancing power supply and demand is performed in a multi-stage, complex process. It starts from bilateral contracts on the forward markets. As we approach to the time of delivery, better forecasts are available. Depending on the remaining time of delivery, various trade options may come into action. Usually, market entities begin trade on power exchanges. At a certain time before delivery, the cross-border trade may also start (in the implicit or explicit auctions). An exemplary structure of process on electrical energy market is presented in Figure 1.

The final balancing segment of the market is called *balancing market*. In the final step, the real-time balancing of power supply and demand is a task for the System Operator. The balancing market can consist of several segments, and typically it consists of the day-ahead, intra-day and real time markets.

3 Day-Ahead and Intra-day Markets

Let us consider a hypothetical implementation of the electrical energy balancing market that is based on M^3 model. We will focus on day-ahead and intra-day markets. The design and operational rules of the balancing markets are specified by applicable regulations. Here we shall follow the regulations of the polish balancing market [8]. Our purpose is to show that all required operational functionalities can be modeled in an open market design standard, by using M^3 application. The application of M^3 forces to use specific terminology related to general multi-commodity markets. However, this seeming complexity can be hidden to the market entities, who are accustomed to a specific nomenclature. M^3 makes the balancing market model elastic and expressive. The rules of the market can be relatively easily and inexpensively evolved, that would be hard to achieve under any specific, hermetic market design.

There are two segments of the polish balancing market, day ahead and intra-day, where trade is organized in sessions. Each session begins with collecting the market data, including submitting the offers by market entities. After that, market clearing process is performed. Finally, there is a settlement phase. This straight mechanism naturally fits into M^3 concept of network of elementary balancing processes. Data flow horizontally through consecutive elementary balancing processes, but also vertically, that is, from day-ahead market processes to intra-day market process and in the opposite way (see Figure 2).

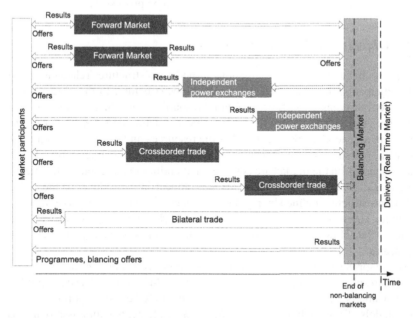

Fig. 1 Exemplary structure of electricity market

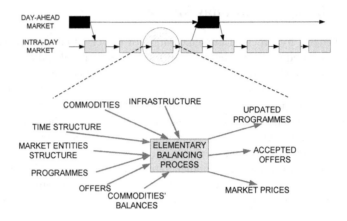

Fig. 2 Segments of balancing market as the network of elementary balancing processes

Below we shall also consider some technical aspects of M^3 application to the real electrical energy balancing market. We shall estimate the volume of data in case of M3-XML as a main data source. Obviously not all data should be stored as a M3-XML files in real applications. However, in the case of research and development, it could be useful to have all data case in files which could be easily shared with others. We consider the memory requirements for storing the data, assuming that data capable to be stored in operating memory can be processed.

3.1   Objects and Entities Structure

The polish balancing market regulations provide two structures related to the market participants: the object structure and entities structure. In the object structure there are four types of objects: generating resources, consuming resources, energy trays and external system resources. The object structure describes rather technical aspects in a general scope, e.g. whether the resource can be controlled centrally by a system, or which resources can be aggregated into one market entity. The contracting parties on the market are considered as entities that are related to objects. In particular, a controllable generating resource appears as a producer on the market.

Entities structure defined by regulations can be modeled directly as a M^3 entities structure. There are no relationships between market entities, thus the M^3 entities model is flat. However, more complex relations can be modeled, such as entities aggregates.

The mapping is presented in Table 1. The producers and consumers which are allowed to submit balancing offers are mapped into M^3 entities. The other producers and consumers that do not participate in offering may be also important for power flow models. They are mapped into infrastructure elements with possibly non-zero divergence or as dummy M^3 entities that are not gaming. A cross-system exchange point can be modeled in a similar way.

Table 1 Object and entities structure mapping into elements of M³

Object structure	Entities structure	Mapping into M³ model
generating resource	producer	a *market entity* (with additional attributes) or infrastructure element element (with additional attributes) or dummy *market entity*
consuming resource	consumer	as above
	-	as above
energy tray	producer/consumer	as above
	-	as above
cross-system exchange point	-	particular kind of *market entity*

Entity types defined in regulations of polish balancing markets can be mapped into three kinds of M³ market entities. Definition of market entities kinds takes negligibly small memory. According to the current regulations, an entity of producer kind has about 71 attributes (including reserves parameters for different kinds of reserves, parameters related to start-ups, forbidden generation regions, other technical constraints and parameters), an entity of consumer kind or balancing entity has about 5 attributes. Figure 3 presents memory size of M3-XML file describing entity structure assuming that average size of attribute definition is 100 characters and there is 25% of producers, 25% of consumers and 50% of balancing entities. The size of xml file cannot be ignored, but it should not be a problem for processing.

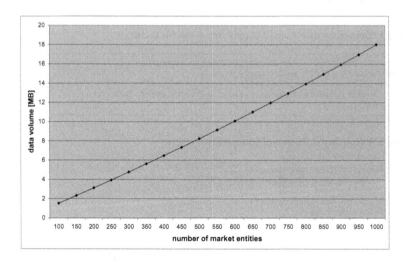

Fig. 3 Estimated size of M3-XML document defining entities structure

3.2   Commodities Structure

Regulations of the polish balancing market provide two main kinds of commodities: market and non-market commodities. There are two families of the market commodities: electrical energy and reserves. The market commodities can be directly mapped into M^3 commodities. Availability of each commodity is defined by relation to infrastructure element and by relation to time slots. The relationship between commodities are not modeled directly, but they arise from more complex infrastructure and calendar models. For instance, on the day-ahead market there is commodity for the electrical energy related to a given hour and 4 commodities defined for each 15-minute slots on the intra-day market. Electrical energy can be traded as a nodal energy (related to a given node of the detailed infrastructure model) or as a zonal energy (related to a subset of nodes of the detailed infrastructure model). This zonal energy must be linked to a virtual node of the infrastructure model.

Finally, other relations may be modeled by using the grouping offers mechanism. For instance, there are several types of reserves of different quality, and demand for a reserve of a given quality can be also satisfied by reserve of better quality. This commodities substitutions can be modeled as a particular grouping offer.

Listing 1 presents exemplary definition of commodity of type op:Electric-Energy which is available at node op:node2 in time element referenced by op:theOnlyPeriod. Exemplary definition of tertiary reserve (the slowest reserve supplied by active or quick-start power plants) is presented in Listing 2. The reserve is a localized (at node op:nodeA) tertiary reserve (kind op:Reserve)). Additional parameters op:reserveDirection indicate that the reserve can only be used to increase power generation.

Listing 1 Exemplary definition of commodity – electrical energy

```
<m3:Commodity id="op:el-2" dref="op:ElectricEnergy"
     minBalance="0.00" maxBalance="0.00">
  <m3:availableAt ref="op:node2"/>
  <m3:CalendarScheduledCommodity ref="op:theOnlyPeriod"/>
</m3:Commodity>
```

Listing 2 Exemplary definition of commodity – reserve

```
<m3:Commodity dref="op:Reserve" id="op:tru-A"
       minBalance="0" maxBalance="INF">
  <m3:description>
    Localized Tertiary Reserve Up - in node A
  </m3:description>
  <m3:parameter dref="op:reserveActivationType">
    Tertiary
  </m3:parameter>
  <m3:parameter dref="op:reserveDirection">Up</m3:parameter>
  <m3:availableAt ref="op:nodeA" />
  <m3:CalendarScheduledCommodity ref="op:theOnlyPeriod" />
</m3:Commodity>
```

Definition of a single commodity does not require a significant amount of memory. However, there may be a considerable number of commodities, since each commodity is related to a given time slot and a given localization. Assuming 200 price nodes, 624 time slots, 4 types of commodities, average size of each attribute definition as 500 characters, the total size of commodities definition is about 60MB.

Commodities are related to time structure, but length of the calendar is relatively small. Assuming two days balancing period, hourly quanta on day-ahead market and 5-minutes quanta on intra-day market and average size of single quant definition equal 200 characters, the size of M3-XML calendar definition is about 11KB.

3.3   Other Balancing Market Elements

There are four categories of market data in regulations of the polish electrical energy balancing market. In the first category there are data variable for each elementary balancing process. These are offers data directly related to commodities structure. All data within this category can be showed to the market in M^3 offers.

Data that is not directly related to commodities structure and which varies more rarely, constitute a second group of market data. They include technical and cost data, like minimal and maximal power output, start-up characteristics, start-up costs, and so on. Data within this category are presented in M^3 model as attributes of the market entities definitions.

Third category of market data are bilateral agreements. Elementary balancing task must be fed with information about bilateral agreements, that is, a total volume of sold or bought energy for each entity. Aggregated volumes of bilateral agreements can be submitted as M^3 programmes.

The last category of market data includes demand forecasts. The forecasts in M^3 model can be represented in two ways: as a balance of each commodity, or by adding virtual entities that represent offers "buy at any price" for a volume representing the forecast.

Another required important data is a model of the transmission network. Balancing processes are based on so called full network model which is detailed model of the power grid. The full network model includes physical network nodes and points of energy delivery (PDE). Some of physical nodes are also marked as price nodes.

Other data needed to perform balancing task are related to market power mitigation mechanisms. The data required by practical market power mitigation mechanisms include following data categories:

- capital groups – they can be modeled as an enhancement of M^3 market entities model with definitions of virtual entities;
- default offers - presented in M^3 as simple offers with additional attribute pointing their role;
- must-run contracts - presented in M^3 as offers;
- competitive network constraints - modeled in M^3 as a additional parameters for an infrastructure elements (e.g. capacity of transmission line).

M3-XML definitions of network infrastructure consists of definition of types of networks and networks definitions. Definition of `<networkKind>` is negligibly small. Estimation of memory requirements is based on the following assumptions.

- Number of lines is equal 1.21 per number of nodes (assumption based on ratio resulting from transmission network models of Polish grid availably in MAT-POWER package [9]).
- Definition of a node consists of following attributes: busBaseVoltage, busMin-Voltage, busMaxVoltage, busShuntSusceptance, busShuntConductance and it takes amout 500 characters.
- Definition of an arc consists of the following attributes: start, end, arcResistance, arcReactance, arcTotalSusceptance, arcLongTermRating, arcShortTerm-Rating, arcEmergencyRating and it takes about 1000 characters.

Exemplary definition of physical node and transmission line are presented in Listings 3 and 4 respectively.

Listing 3 Exemplary definition of node

```
<m3:node id="op:nodeA" dref="op:Bus">
    <m3:name>Node 1</m3:name>
    <m3:parameter dref="op:busBaseVoltage">220</m3:parameter>
    <m3:parameter dref="op:busMinVoltage">0.95</m3:parameter>
    <m3:parameter dref="op:busMaxVoltage">1.11</m3:parameter>
    <m3:parameter dref="op:busShuntSusceptance">0</m3:parameter>
    <m3:parameter dref="op:busShuntConductance">0</m3:parameter>
</m3:node>
```

Listing 4 Exemplary definition of transmission line

```
<m3:arc id="op:arcL1" dref="op:ACLine">
        <m3:name>Line L1</m3:name>
        <m3:parameter dref="op:arcResistance">0.0</m3:parameter>
        <m3:parameter dref="op:arcReactance">20.0</m3:parameter>
        <m3:parameter dref="op:arcTotalSusceptance">
          0.0
        </m3:parameter>
        <m3:parameter dref="op:arcLongTermRating">50</m3:parameter>
        <m3:parameter dref="op:arcShortTermRating">0</m3:parameter>
        <m3:parameter dref="op:arcEmergencyRating">0</m3:parameter>
        <m3:predecessor ref="op:nodeA"/>
        <m3:successor ref="op:nodeB"/>
</m3:arc>
```

In Listing 5 there is definition of virtual node which aggregate nodes `op:nodeA` and `op:nodeB`.

Exemplary definition of virtual line is presented in Listing 6. For physical elements of the network all parameters are known. For virtual lines only the capacity is defined as a simple parameter. There are also complex parameters, so called Power Transfer Distribution Factors factors (PTDFs).

Data volumes of programmes are relatively small. For 200 market entities, 4 commodity kinds, and 48-stages horizon, estimated size of programmes data volume is about 18MB (assuming 500 characters per market entity and commodity kind).

Listing 5 Exemplary definition of virtual node

```
<m3:node id="op:copperPlate" dref="op:Zone">
     <m3:aggregates ref="op:nodeA"/>
     <m3:aggregates ref="op:nodeB"/>
</m3:node>
```

Listing 6 Exemplary definition of virtual line

```
<m3:arc id="op:arc-A" dref="op:AggregatedLine">
  <m3:description>
    Connections between DE (North) and PL zones
  </m3:description>
  <m3:parameter dref="op:aggregatedArcCapacity">1000</m3:parameter>
  <m3:complexParameter dref="op:PTDF">
    <m3:complexValue>
      <m3:complexValue dref="op:PTDFValuesForSource">
          <m3:reference dref="op:sourceNode" ref="op:zoneDEN"/>
          <m3:complexValue dref="op:singlePTDFValue">
            <m3:reference dref="op:sinkNode" ref="op:zonePL"/>
            <m3:value dref="op:ValueofPTDF">0.70</m3:value>
          </m3:complexValue>
          <m3:complexValue dref="op:singlePTDFValue">
            <m3:reference dref="op:sinkNode" ref="op:zoneSL"/>
            <m3:value dref="op:ValueofPTDF">0.66</m3:value>
          </m3:complexValue>
        <!-- ... -->

  </m3:complexParameter>
  <m3:predecessor ref="op:zoneDEN"/>
  <m3:successor ref="op:zonePL"/>
</m3:arc>
```

Last important data category are offers. Under some simplifications we can assume that each market entity is allowed to submit 10 offers in a single elementary balancing process. Assuming 48 commodity kinds on day-ahead market and 24 commodity kinds on intra-day market, 500 characters per single definition, the estimated size of offers is about 45MB and 23MB for each market respectively.

All assumptions are overestimated, thus in practice the sizes of M3-XML documents can be significantly smaller. For example, in data cases repository availably on http://www.openm3.org, average size of arc definition is 507 characters (1000 characters assumed), average size of node definition is 350 characters (500 assumed). Also estimations resulting from polish market regulations, are much lower, e.g. number of producer/consumer attributes is 71/5 respectively (100 for producer kind and 20 for consumer kind assumed), number of time slots in a single balancing process is 24 (48 assumed).

Sizes of M3-XML cannot be ignored, however at current achievements in processing XML files it should not rise any barriers. Figure 4 presents volumes of all data under assumption of 200 market entities, 2400 nodes in grid, 48-stages planning horizon, and PTDFs for 10% of all lines.

Since many of data has some regularity, they can be notated in a much more compact way. Some attempts, including M3-XML extension can be found in [1].

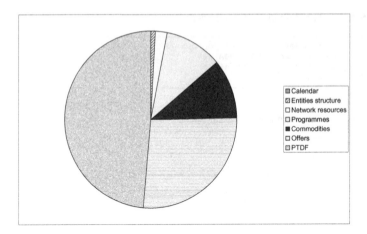

Fig. 4 Estimated data volumes for each categories of data of single elementary balancing process

4 Development of Balancing Markets with M^3 Model

Electrical energy markets are under continuous development. In this section we present some possible evolutionary steps when using M^3 model.

4.1 Aggregations

Two types of aggregations are discussed. The first is the aggregation of physical nodes into price zones (price hubs). The open question is whether zonal pricing is better than nodal pricing and if so, how zones should be created. M^3 models models so called virtual network with virtual nodes that may represent zones. Several physical and virtual networks can be used to reflect different context. For power flow computations a detailed network model is used. For pricing issues virtual network can be used to model zones. One application of price hubs is trading of energy on so called "copper plate" market, i.e. without localizations, – electrical energy is related to an aggregated node that is representing a whole system.

Another issue is the aggregation of entities. In M^3 model any aggregation of entities can be represented through market entities structure, where the relation between entities can be defined. Various virtual entities, such as balancing groups, can be defined. Aggregation can be done in a multi-stage manner – some virtual entities can be further aggregated into a more aggregated entity. This mechanism is sufficient to provide required data to elementary balancing processes in various scenarios.

4.2   Elastic Demand

Let us consider a consumer who must buy energy needed to carry out production process. We consider fifteen-minute stages in balancing horizon. Suppose that a participant must execute a production cycle, which lasts an hour and is characterized by a variable energy demand. Energy demand is shown in Picture 5 and Table 2. The consumer may not change start date of the production cycle, while he may cancel production in this period. He can not carry out a only part of the production process. The participant decides not to purchase energy, if the price of the set 1.8 MWh of electricity exceeds 360$.

Table 2 Energy demand

Stage	1	2	3	4
Demand[kWh]	400	400	200	800

The customer would like to:

- buy the same volume of energy in the first and the second stage,
- in third stage buy a half of energy in the first stage,
- in the fourth stage buy two times more energy than in the first stage.

To model such preferences, a bundle offer can be used. The consumer should submit bundle offer with parameters described in Table 3. Since the maximal price is 360$ and define maximal volume as a 400 units, then unit price is 0.9$. Share factors directly represents consumer preferences.

In this case a bundle offer mechanism enables the entity to show preferences more accurately than in a traditional offering mechanisms, in which offering for each stage should be considered separately.

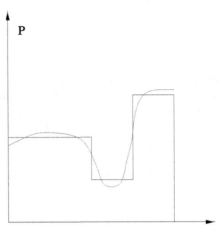

Fig. 5 Energy demand   T

Table 3 Parameters of bundle offer

Price	0.9
Volume	$< 0,400 >$
Share factor for stage 1	-1
Share factor for stage 2	-1
Share factor for stage 3	-0.5
Share factor for stage 4	-2

4.3 Start-Up Service as a Market Commodity

Let us consider the start-up of generators treated as a service on the market. For simplicity, we will focus on one type of start-up only. Let S be a start-up price in an offer of a generator. To satisfy the relations between start-up service and produced energy we need to add a new commodity – generator activity. The generator is not submitting any offer of its activity, but the activity results from decisions of accepting start-up services and energy offers. Table 4 presents the commodities, where h is the index of time period, p_h is accepted volume of energy, binary v_h is 1 if the generator is on and 0 if the generator is off; and w_h points (with value 1) time period where generator starts to work after start-up.

Table 4 Commodities definition

Kind	Energy	Activity	Start-up
Symbol	p_h	v_h	w_h
Domain	$\{0\} \cup [0, p_h^{max}]$	$\{0,1\}$	$\{0,1\}$
Price	A_h	0	S

In balancing models additional constraints should appear:

$$p_h \leq v_h M, \quad \forall h \in H \tag{1}$$

$$v_h - v_{h-1} \leq w_h, \quad \forall h \in H \tag{2}$$

where H is a set of period in horizon and M is sufficiently large. To provide balancing mechanism with information on above constraints, the generator can submit a grouping offer. Parameters of appropriate offer are presented in Table 5. We assume that generator has submitted simple offers for energy in consecutive periods with offer prices s_{h-1}, s_h and s_{h+1}, and he has submitted offers for start-up with prices w_{h-1}, w_h and w_{h+1}. The grouping offer provides definitions for constraints binding simple offers which reflect inequalities (1)-(2).

Table 5 Grouping offer with start-up costs

type	energy			activity			start-up				
price	p_{h-1} s_{h-1}	p_h s_h	p_{h+1} s_{h+1}	v_{h-1} 0	v_h 0	v_{h+1} 0	w_{h-1} S	w_h S	w_{h+1} S		
	1			$-M$						$\leq$	0
		1			$-M$					$\leq$	0
			1			$-M$				$\leq$	0
					...					$\leq$	0
				-1	1		-1			$\leq$	0
					-1	1			-1	$\leq$	0
					...					$\leq$	0

5 Real Time Segment

In the electricity market system there is a need for market mechanisms which would be able to balance demand and production in time close to the delivery (realization time) [5]. This mechanism is called the *real time market* and is usually organized as a separate and final segment of the balancing market. The main aim of this segment is to give to market participants possibility of trading in so-called real time as well as to allow the System Operator to close the system balance respecting all technical constraints using market mechanisms [6]. This approach may increase system efficiency and its transparency.

Real time market gives new possibilities both to the market participants and to the System Operator. But this segment also raises some new requirements for balancing and pricing mechanisms as well as for data exchange mechanisms. In the following sections we present these requirements and analyze possibilities of M³ usage on this segment of the market.

5.1 Communication Efficiency

Due to very strict time limits, real time market mechanisms have to be very efficient. On many existing power markets real time balancing process are repeated in every 5 minutes, and mechanism has to give a response within this time. One of the elements, which may be crucial to meeting this requirements, is the efficiency of the communication mechanism. Before each iteration mechanism has to be supplied with actual offers and data about system state and after each balancing process, new work programmes have to be send to the market participants.

Using M³ we can easily meet such efficiency requirements. In experiments made so far [4] time used for communication between autonomous agents was in order of a few milliseconds, which is small enough comparing to balancing frequency required on the electric power market.

5.2   System Operator – Central Agent

Real time segments of the existing power markets are usually organized as centralized markets, which are controlled by the Independent System Operator (ISO) ot Transmission System Operator (TSO). Severely limited operation time does not allow for time consuming negotiations and potential trade partner searching, and, what is more crucial, balancing results have to meet all transmission network constraints. Efficient communication environment could alleviate first restriction, while meeting network constraints in reasonable time could be very hard in the distributed environment. To meet all constraints there is a need for a central unit, which has full knowledge about system and its current state and is entitled to control each of its part, including requirements of all market participants.

5.2.1   Requirements

System Operator as a central market agent should meet some requirements:

- SO should be well known to all market participants;
- SO should have full and detailed knowledge about the power system and about all agents;
- all agents should be able to communicate easily with SO;
- SO should be able to communicate simultaneously (and effectively) with many agents;
- SO should be able to effectively process many communicates from other agents (we expect that in some time periods most of agents will tray to communicate with SO).

M^3 meets all requirements towards System Operator. In the centralized communication model ([3], see also Chapter *Communication models used in the context of multi-commodity trade*) there is an assumption that there exist the central agent, which has full knowledge about the system. All other agents have to register to SO, and during registration process they obtain detailed information about communication with SO. During registration proces SO may also obtain detailed information about each agent. This data are stored using the M^3 element <entities>. Using this element of M^3 System Operator may gather all unit technical parameters like: start-up characteristics, minimal and maximal production levels, possible speed of increase and decrease production, etc. System Operator may use the <entities> element also for storing other information like ownership structure, reference production costs, etc.

Moreover, System Operator may store all information about transmission network (structure and constraints) using the M^3 <networks> and <networkskind> elements. Detailed possibilities of infrastructure modeling could be found in [2].

Conducted experiments [4] show, that in the centralized market structure total time needed for communication in one iteration of market processes was in the order of seconds. It means, that it is significantly smaller than time which is required for market balancing, and M^3 may be efficiently used as data and communication model for energy real time markets.

5.3   Trading without Human Surveillance

Relatively high frequency of real time markets processes causes, that transactions conducted by automatic software agents could be a desirable solution. It may be more convenient for market participants, but also more efficient, because automatic agents could better monitor market state and conduct transaction much faster than humans. Possibility of conducting transactions by automatic agents rise some additional requirements towards communication model, like:

- well defined communication model;
- well defined exceptions handling;
- possibility of automatic data correctness and reasonableness verification;
- possibility of exact market participant's preferences mapping;
- possibility of agent's strategy definition;
- communication safety: authorization / confidentiality/ non-repudation / integrity / acknowledgment, ...

The M^3 provides well defined communication model, which is suitable for automatic trading on most types of markets as well centralized as distributed ones [3]. Communication model encompasses two phase commit protocol, which make possible transaction acknowledgment in the distributed environment.

The M^3 data schema allows for data correctness and reasonableness verification on a few levels. First correctness could be checked during XML file validation using defined M^3 XML Schemas. If file is correct from the XML point of view it could be verified from the data schema point of view. There may be checked if required parameters for each elements are defined and if they have appropriate type and range. Finally correctness and reasonableness of data cold be checked against other data from other sources. System Operator may check if offered volume coincides with maximal possible generation of this unit or if submitted bilateral contract is confirmed by both sides of the transaction.

Using a wide range of offers defined in M^3 user may define his preferences. There may be used either simple offers for one commodity, or integrated offers, which allow to define offer or demand for a bundle of goods (e.g electric power in few consecutive hours, or electric power together with transmission rights) and grouping offers, which may be used to define more sophisticated requirements. Grouping offers may be used e.g. to limit possible generation change between two consecutive hours (so called ramp constraint) or to tie offers for electric energy and regulation reserves, which are limited by a common resource – maximal generation capabilities.

An example of grouping offer, which defines a ramp constraint is presented in the Listing 7.

Listing 7 An example of grouping offer

```
<m3:GroupingOffer groupingFunction="m3:linear-grouping"
      id="op:ogr1_1_2">
   <m3:description>Ramp grouping offer</m3:description>
   <m3:offeredBy ref="op:gen1" />
   <m3:groups>
     <m3:groupedCommodity ref="op:r0102g1" />
     <m3:groupedOffer coefficientValue="1"
       ref="op:offer1_p2" />
   </m3:groups>
   <m3:groups>
     <m3:groupedCommodity ref="op:r0102g1" />
     <m3:groupedOffer coefficientValue="-1"
       ref="op:offer1_p1" />
   </m3:groups>
 </m3:GroupingOffer>
```

The offer `op:ogr1_1_2` groups two simple offers: `op:offer1_p1` and `op:offer1_p2`, which are respectively offers for selling energy in two consecutive periods 1 and 2 submitted by entity `op:gen1`. The coefficient value for offer `op:offer1_p1` is -1, whereas for offer `op:offer1_p2` is equal to 1. The offers are bundled with the same commodity `op:r0102g1`. There is used a linear grouping function, therefore this offer limits the difference: `op:offer1_p2` $-$ `op:offer1_p1`, which defines generation increase between periods 1 and 2 for entity `op:gen1`. The `op:r0102g1` is a virtual commodity defined as in Listing 8.

Listing 8 An example of grouping offer

```
<m3:Commodity dref="op:Ramp" id="op:r0102g1"
      maxBalance="50"  minBalance="-100">
   <m3:description>Ramp between period 1 and 2 for gen 1
      </m3:description>
   <m3:VirtualCommodity />
   <m3:availableAt ref="op:node1" />
   <m3:CalendarScheduledCommodity ref="op:P01260507" />
 </m3:Commodity>
```

The balances of virtual commodity limit maximal difference between generation level of generator `op:gen1` in consecutive periods to 50 MWh in case of generation increase and to 100 MWh in case of generation decrease (negative value of the expression: `op:offer1_p2` $-$ `op:offer1_p1`) . Whole content of all files in this example could be found at the M^3 website [7].

All communication safety issues may be ensured using well known asymmetric cryptography methods. Detailed approach has to respect specific demands of the particular system and this problem will be not considered further in this work.

Further research may concern exceptions handling in communication model and issues concerning detailed information about offer rejection. In the agent system information about reasons of offer rejection, provided in a form which may be processed by automatic agents could be very useful. Agent may make an attempt to correct earlier submitted offers or at least may know who should be informed about the problem.

5.4 Multi-stage Balancing

As it was described in Section 2 the power market, like many others infrastructure markets, consists of many segments. Therefore balancing mechanisms have to take into consideration contracts conducted in previous market processes. For that reason the data model ought to allow for storage information concerning all so-far conducted contracts and work programmes, which results from them. It is especially important on the real time segment, because due to short interval between trading and delivery time, as well as continuous character of the market planing horizon, consecutive processes often overlap. Consequently each real time market process has to take into consideration not only outcome of previous segments but also results of preceding real time market processes.

In the M^3 there is defined the <programme> element, which could be used to store information about current state of the whole market or its part. It may be used both by System Operator and market entities. Using this element market entities may store information about contracts, which they have conducted on the previous market segments and send this information to the System Operator. While System Operator may use this element to create complete picture of the system state before start of each balancing process.

The <programme> element may be used for storing data concerning physical contracts for commodities which are traded on the balancing market as well as various range of derivatives, like: financial contracts, options, transmission rights, etc. Consequently each market entity could submit offer for change of its earlier established work programme.

5.5 Infrastructure and Time Mapping

As it was mentioned earlier, on the real time segment there is a need for exact representation of the system infrastructure in the data model. Due to rigid transmission constraints, commodities has to be defined in a specific place and time, there ought to be additional possibility of tying commodities to the elements of infrastructure as well as to the specific time periods.

The infrastructure data model should also allow to define various range of constraints. They may be defined at the whole system level (system constraints), like e.g. line transmission capabilities and safety requirements, as well as on the single market entity level (individual constraints), like e.g start-up time, generation capabilities and ramp constraint.

In section 5.2 the possibilities of mapping system infrastructure into M^3 were described. Each commodity defined on a market could be related to any node or line defined in the infrastructure model, using the M^3 element <availableAt>.

Each commodity could be also related to the time structure using the M^3 element `<CalendarScheduledCommodity>`.

In M^3 the time structure could be hierarchical, which is important from the real time markets points of view. In real time market the basic time unit is usually significantly smaller than in the other segments. In many real-world power markets in the real-time segments power is traded within 5 or 15 minutes periods, while in the other segments the basic period is 1 hour. Thanks to a hierarchical time structure the commodities may be automatically transformed between market segments.

Individual entities constraints may be defined in M^3 using grouping offers (see Section 5.3), which are flexible enough to define each requirements and dependencies, which can be expressed as a mathematical formula. Basic system constraints in M^3 are strictly connected with a infrastructure model and could be defined within definition of infrastructure elements. More sophisticated system constraints, like e.g. safety requirements, could be also expressed using grouping offers, which are submitted by the System Operator.

5.6 M^3 *Usage on the Real Time Markets*

Real time market segment raises some new requirements regarding data exchange mechanisms. The M^3 model meets most of these demands and could be effectively used to store and exchange data on such markets, especially on the real time power market. Data model is expressively enough to store all necessary data and to map all demands. Communication model is efficient enough to trade effectively and safely on the real time market. Some system specific and additional requirements may require further researches, which may enrich so far obtained functionality, but this should not limit possibilities of the M^3 usage.

6 Summary

M^3 standard may be a convenient design solution for the electricity markets. It meets all requirements specific to this economic sector and allows for gradual improvements of trading possibilities by using advanced multi-commodity trade mechanisms. M^3 may be used in all electricity market segments, despite the fact that they may be significantly diverse. The common data and communication model applied to all segments may substantially facilitate trading in electricity markets both for market entities as well as to the System Operator. What is more important, due to standard openness, its flexibility ad expressiveness, it may be used in many different implementations of the power market and does not need to be changed during market restructuring. Therefore, any changes in the market balancing mechanisms does not have to entail exchange of all IT systems used by market entities and the System Operator.

References

1. Kacprzak, P., Kaleta, M., Pałka, P., Smolira, K., Toczyłowski, E., Traczyk, T.: Notation methods for large volume regular data in complex electronic trade problems. In: Górski, A. (ed.) Information Systems Architecture and Technology, OWPW, Wrocław (2010)
2. Kacprzak, P., Kaleta, M., Pałka, P., Smolira, K., Toczyłowski, E., Traczyk, T.: Data model for an open Multi-commodity Market Model M³. In: Kozielski, S., et al. (eds.) Databases. New Technologies, pp. 289–300. WKŁ, Warszawa (2007)
3. Kacprzak, P., Kaleta, M., Pałka, P., Smolira, K., Toczyłowski, E., Traczyk, T.: Communication Model for M³ – Open Multi-commodity Market Data Model. In: Proc. 2nd National Scientific Conference on Data Processing Technologies KKNTPD 2007, pp. 139–150. Poznań (2007)
4. Pałka, P., Całka, M., Kaleta, M., Toczyłowski, E., Traczyk, T.: Design and Java implementation of the multi-agent platform for multi-commodity exchange. In: Proc. 3rd National Scientific Conference on Data Processing Technologies KKNTPD 2010, pp. 184–196. Poznań (2010)
5. Smolira, K., Toczyłowski, E.: Real-time Market Mechanisms for Control in Distributed Networks. In: 12th IEEE International Conference on Methods and Models in Automation and Robotics, pp. 281-285 (2006)
6. Smolira, K.: Analysis of the real-time markets balancing mechanisms. PhD Thesis. Warsaw University of Technology, Warszawa (2008) (in Polish)
7. M³ Web Site, http://www.openm3.org/
8. PSE Operator S.A., Instruction of Transmission System Operation and Maintenance (in Polish), http://www.pse-operator.pl
9. MATPOWER Web Site, http://www.pserc.cornell.edu/matpower/

A Market for Pollution Emission Permits with Low Accuracy of Emission Estimates

Zbigniew Nahorski and Joanna Horabik

Abstract. Uncertainties of pollution inventories are often high due to low precision of emission quantity assessments for many emitting sources. A good example is emission of greenhouse gases, where uncertainty of some sources may be as high as 40-100%, while uncertainty of other sources is as low as 2-3%. This discrepancy in uncertainty should be accounted for in compliance as well as in emissions trading, because the traded commodities have different quality. The compliance and emissions trading rules have been discussed in earlier papers by the present authors [14, 15, 16, 17]. In this chapter we focus on presentation of the idea of a market for emissions with so highly scattered uncertainties.

1 Introduction

A market in emission permits is closely connected with basic markets for goods produced by the emitting entities. As these entities may produce very different goods, the market in emission permits binds together many markets, possibly coupled mutually, and moreover possibly involved with markets in derivatives. Thus, the entire net of these markets forms a large and extremely complicated multi-commodity structure. Analysis of this structure is out of scope of this chapter. We concentrate here only on an emission permits market, where seemingly only one commodity is traded. But in fact it is a very inhomogenous one. Although this market could be perhaps designed as a number of interconnected markets for more homogenous goods, the propriety of this specific good allows us to propose a specific market, where the inhomegeneity of the good is compensated for in the trading rules.

Zbigniew Nahorski
Systems Research Institute, Polish Academy of Sciences
e-mail: Zbigniew.Nahorski@ibspan.waw.pl

Joanna Horabik
Systems Research Institute, Polish Academy of Sciences
e-mail: Joanna.Horabik@ibspan.waw.pl

M. Kaleta & T. Traczyk (Eds.): Modeling Multi-commodity Trade, AISC 121, pp. 149–163.
springerlink.com © Springer-Verlag Berlin Heidelberg 2012

Emissions of environmental pollutions are often too difficult to be measured directly. Therefore, they are calculated indirectly by inventorying activities causing emissions, and estimating their influence on the final emissions. These estimates are due to different errors, like low precision of amount of source materials used, their quality, insufficient knowledge of processes emitting pollutions, etc. A good example of this kind of procedure is emission of greenhouse gases. Although emissions of carbon dioxide from large professional power plants can be estimated with quite satisfactory uncertainty of 2-5%, uncertainty of methane emissions may reach up to 40%, and uncertainty of nitrous oxide inventories is even more than 100%. There arises the problem how to evaluate compliance with imposed limits on such uncertain emissions, and how to trade emission permits.

The problem how to treat uncertain greenhouse gases emissions has been discussed already for some years [3, 6, 7, 8, 9, 10, 11, 22]. In particular, some ways of solving the permit trading were proposed. For example, in [12] it was proposed to exclude uncertain emissions from trade, or to trade emissions of similar uncertainty on separate markets.

In this chapter we shall focus on so-called undershooting approach [4, 13, 17, 16, 14]. Our aim is to outline the market for emission permits with highly diversified emission uncertainties.

In Section 2 we discuss the problem of evaluation of compliance for uncertain emissions with different uncertainty distributions. In Section 3 we deal with the asymmetric interval uncertainty and we derive conditions for checking compliance in such a case. In Section 4 we present so-called effective emissions, which can be traded directly, without taking into account emission uncertainty. In Section 5 the market with effective emissions as additional instruments is discussed, while Section 6 considers the market with solely effective emissions used in trading. The market rules are given and market properties are derived. Section 7 concludes.

2   Uncertainty and Evaluation of Compliance

When emisions are known exactly, for a given upper limit L imposed on emissions, the actual emissions x must satisfy the condition

$$x \leq L \tag{1}$$

In some cases, a reduction of inventory is given in percents. Denoting x_c as an emission inventory at the end of reduction period, x_b as an emission inventory in the beginning of the reduction period, and ρ as a required fraction of emission reduction, then the compliance condition is $x_c \leq \rho x_b$. It can be transformed to $x_c - \rho x_b \leq 0$. This way, the condition has the form (1), with $x = x_c - \rho x_b$ and $L = 0$. To simplify argumentation, only the condition of type (1) will be considered in the sequel.

The problem arises when the emissions x are not known with a satisfactory accuracy. For example, this is the case when emissions are estimated from an approximate inventory, as it is for emissions of greenhouse gases. In order to highlight uncertainty of such an estimate, this value is denoted in the sequel by $\hat{x}$. Moreover, distributions of uncertainty for different gases as well as for national inventories may be asymmetric [19, 23].

For better illustration of the problems related to dealing with uncertain emissions, let us look at simplified distributions $\mu(x)$ of two inventories, A and B, presented in Figure 1, shifted to zero at the limit L, often called the target. The calculated inventories $\hat{x}$, called here the nominal values, correspond to the highest values of the distributions. Taking into account only the nominal values, the party with inventory A fulfills the emission condition (1), while the party with inventory B does not. However, if the distributions are interpreted as the probability distributions, one can see that the probability that the real inventory does not fulfill the limit (the area under the distribution for the positive values of x) is higher than the probability that it does not (the area for the negative values of x). One may ask the question which criterion is better suited to order uncertain inventories.

The impression that the inventory A is not necessarily better that B is actually questioned by many techniques used to compare uncertain values, see [5]. Here we mention few of them.

The most elementary technique of ordering uncertain values is based on *the mean value and the variance* (MV). According to this technique, the smaller the mean value and the variance are, the better the inventory is. For the example presented in Figure 2, the respective values are depicted in Table 1. Although the nominal value of the inventory A is smaller than that of B, the mean value of A is greater than the mean value of B. The same is true for the standard deviations. Thus, even this simple criterion shows that an inventory of the party B should be considered smaller than an inventory of the party A. This is contrary to the result for nominal values, which ignores uncertainty.

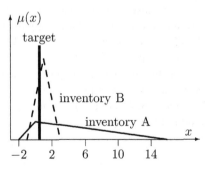

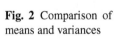

Fig. 1 Distributions of two inventories considered in the text

Fig. 2 Comparison of means and variances

Table 1 Criteria values for comparison of inventories A and B

Method	Criterion value for A	Criterion value for B	Inventory chosen
MV	$m_A = 4$ $\sigma_A = 16\frac{1}{3}$	$m_B = 1$ $\sigma_B = \frac{2}{3}$	B
MSV	$s_{SA}^2 = 13.45$	$s_{SB}^2 = 0.35$	B
CP	$crp_A = \frac{8}{9}$	$crp_B = \frac{7}{8}$	B
risk	$c_{critA} = 10.6$	$c_{critB} = 2.1$	B

A large group of techniques uses the notion of *critical probability* (CP), proposed already in 1952, [20]. The methods in this group require knowledge of respective probability distribution $\mu(x)$. The measure used to compare inventories is the probability of surpassing the target L

$$crp = \int_L^\infty \mu(x)dx \qquad (2)$$

A smaller value of crp indicates better inventory. According to Table 1, again, an inventory of the party B is evaluated as the smaller one.

In other related methods, as *the Baumol's risk measure* and *the value at risk* (VaR), the probability of inventory x to be above a critical value x_{crit} is fixed, and then the value x_{crit} is calculated, see Figure 3. Without going into details, an inventory is smaller when x_{crit} is smaller. In our example, fixing probability to 0.1, the inventory B is chosen as the smaller one.

In conclusion, decision about fulfillment of an obligation, which is based on deterministic (nominal value) comparison of an inventory with a target, contradicts the already existing scientific knowledge on ordering uncertain projects using the stochastic approach.

A technique similar in spirit to the CP and risk measures has been proposed to ensure a reliable compliance in the context of greenhouse gases. It is called *undershooting* ([3, 4, 17, 14, 13], and it is illustrated in Figure 4. In this approach, it is required that only a small enough α-th part of an inventory distribution may lie

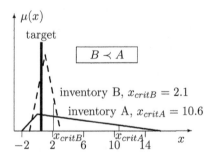

Fig. 3 Calculation of critical values

above a target. This idea, when used for ordering inventories, becomes equivalent to the CP technique.

The undershooting technique is used throughout the chapter to design a market for the emissions with highly diversified uncertainties by introducing comparable quotas which depend on the uncertainty levels.

3 Compliance

Let us denote the lower spread of the uncertainty interval by d^l and the upper spread by d^u. Then, the actual (unknown) emission x is situated in the intervals

$$x \in [\hat{x} - d^l, \hat{x} + d^u]$$

The limit L is known exactly. To be completely sure that a party fulfills the limit, its emission inventory $\hat{x}$ should satisfy the following condition, see Figure 5 (a).

$$\hat{x} + d^u \leq L \tag{3}$$

As the bounds can be quite large, a weaker condition will be used, see [14].

Definition 1. *A party is compliant with the risk α if its emission inventory satisfies the condition*

$$\hat{x} + d^u \leq L + \alpha(d^l + d^u) \tag{4}$$

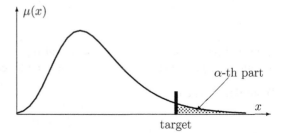

Fig. 4 Illustration of compliance in the undershooting approach

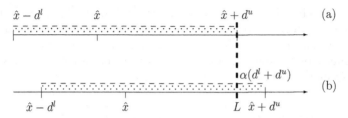

Fig. 5 Full compliance (a) and compliance with risk α (b) in the interval uncertainty approach

Here the risk is understood as a likelihood that a party may not fulfill the agreed obligation regarding the emission limit, due to uncertainty of the emission inventory.

Condition (4) means that the αth part of the party's emission estimate (inventory) uncertainty interval is allowed to lie above the limit L, see Figure 5 (b). After some algebraic manipulations the condition (4) can be also written in the following form

$$\hat{x} + \left[1 - \left(1 + \frac{d^l}{d^u}\right)\alpha\right]d^u \leq L \tag{5}$$

The above condition shows that a part of the upper spread of the uncertainty interval is added to the emission estimate before compliance is checked. This can be also interpreted to mean that an unaccounted emission, due to uncertainty, is included in the condition to reduce the risk of non-compliance. Let us denote by $R^u = d^u/\hat{x}$ and $R^l = d^l/\hat{x}$ the relative upper and lower spreads of the uncertainty intervals, respectively. Denoting the fraction of the unaccounted emission in the emission estimate as

$$u(\alpha) = \left[1 - \left(1 + \frac{d^l}{d^u}\right)\alpha\right]R^u \tag{6}$$

equation (5) can be also written in the form

$$\hat{x}[1 + u(\alpha)] \leq L \tag{7}$$

Definition 2. *We call the left hand side value in (5) or (7) the expanded estimated emission and denote it as*

$$\hat{l} = \hat{x} + \left[1 - \left(1 + \frac{d^l}{d^u}\right)\alpha\right]d^u = \hat{x}[1 + u(\alpha)] \tag{8}$$

Instead of recalculation of emissions, it is also possible to recalculate limits. Thus, we introduce the value, which will be of use in the sequel.

Definition 3. *The value*

$$\tilde{L} = \frac{L}{1 + u(\alpha)}$$

is called the corrected limit.

All possible combinations of emissions and limits: $(\hat{x}, L)$, $(\hat{l}, L)$, $(\hat{x}, \tilde{L})$ and $(\hat{l}, \tilde{L})$, can be used for checking compliance. Any choice will result in particular market rules. However, the most promising are the combinations $(\hat{l}, L)$ and $(\hat{x}, \tilde{L})$, as they provide direct comparisons of, possibly transformed, emissions and limits. These two combinations will be considered in the sequel.

4 Effective Emissions

The above compliance-proving policy can be used to modify the rules of emission trading. The main idea presented in earlier papers [14, 16, 17] involves transferring the uncertainty of the seller's emissions to the buyer's emissions together with the quota of traded emissions, and then including it in the buyer's emission balance.

Let us denote by $\hat{E}^S$ the amount of estimated seller emission allocated for trade. Emission $\hat{E}^S$ is associated with the lower and upper spreads of the uncertainty intervals $\hat{E}^S R^{lS}$ and $\hat{E}^S R^{uS}$, respectively.

Before a transaction the buyer has to satisfy the condition (5), which is reformulated to

$$\hat{x}^B + d^{uB} - (d^{lB} + d^{uB})\alpha \le L^B$$

After buying $\hat{E}^S$ units of emissions from the seller and including the corresponding uncertainty in the formula, the buyer's condition becomes

$$\hat{x}^B - \hat{E}^S + d^{uB} + \hat{E}^S R^{uS} - (d^{uB} + \hat{E}^S R^{uS} + d^{lB} + \hat{E}^S R^{lS})\alpha \le L^B \qquad (9)$$

Two above conditions differ in the value defined below.

Definition 4. *The value*

$$E_{eff} = \hat{E}^S \left\{ 1 - \left[1 - (1 + \frac{d^{lS}}{d^{uS}})\alpha \right] R^{uS} \right\} = \hat{E}^S \left[1 - u^S(\alpha) \right] \qquad (10)$$

is called the effective emission [14].

Note that the effective emission is smaller than the estimated emission. The bigger the relative upper spread of the uncertainty interval of the seller is, the smaller is the effective emission. Effective emissions depend also on the ratio d^{uS}/d^{lS}, and obviously on α.

5   Trading Estimated Emissions versus Trading Effective Emissions

The market for uncertain inventories has been discussed in [1, 2, 14, 16]. It was formulated as an optimization problem with minimization of the sum of costs to achieve the common limit on emissions, subject to compliance with the risk α. This simulation was, however, quite far from the real market conditions. First of all, in the real market, the parties take decisions without knowledge of the cost characteristics of the partner in trading. A rough cost evaluation is done in a process of price negotiation between parties. The idea of the trading prices was introduced in [21], however, prices were only drawn randomly from the feasible region, and they were not negotiated. Some elements of negotiations were used in [18], but uncertainty was not considered there.

Here we discuss organization of the market where uncertainties are taken into account. This is done by using the effective emissions in the trade. The discussion is directed to presenting rules of trading for the individual participant of the market. Some ideas and attempts in this direction were presented already in [14, 16]. Here these ideas are elaborated further and presented in an advanced form. In particular, the combination of emissions and limits $(\hat{x}, \tilde{L})$ is proposed and discussed for the first time.

The basic assumption of this market is that selling/buying emissions is combined with simultaneous transferring of the corresponding uncertainty. The amount of the traded emissions is connected with the effective emissions. We denote the initial values before trade by the subscript $_0$, and those after the transaction number $t \geq 1$ by the subscript $_t$. Let us assume that the amount of $\hat{E}_t^S$ is sold from the seller S to the buyer B. The lower e_t^{lS} and the upper e_t^{uS} uncertainty spreads related with this amount are

$$e_t^{lS} = R_0^{lS}\hat{E}_t^S = \frac{\hat{E}_t^S}{\hat{x}_0^S}d_0^{lS} \qquad e_t^{uS} = R_0^{uS}\hat{E}_t^S = \frac{\hat{E}_t^S}{\hat{x}_0^S}d_0^{uS} \tag{11}$$

where $d_0^{lS} = d^{lS}$, $d_0^{uS} = d^{uS}$, $R_0^{lS} = R^{lS}$ and $\hat{x}_0^S = \hat{x}^S$. Thus, after the transaction we have

$$\hat{x}_t^S = \hat{x}_{t-1}^S + \hat{E}_t^S \qquad \hat{x}_t^B = \hat{x}_{t-1}^B - \hat{E}_t^S \tag{12}$$

with $\hat{x}_0^B = \hat{x}^B$. According to the rules of interval algebra we have

$$d_t^{lS} = d_{t-1}^{lS} + e_t^{lS} \qquad d_t^{uS} = d_{t-1}^{uS} + e_t^{uS} \tag{13}$$

$$d_t^{lB} = d_{t-1}^{lB} + e_t^{lS} \qquad d_t^{uB} = d_{t-1}^{uB} + e_t^{uS} \tag{14}$$

In the usual condition, the seller's estimated emission is distinctly less than the limit, while the buyer's is distinctly higher. Thus, the transaction helps the buyer to achieve his limit.

Above, we consider only two trading parties, and therefore both their estimated emissions change in transactions. Estimated emissions of other parties, besides the ones for the trading parties, do not change after transaction t. This is formally written by taking for them $\hat{E}_t^S = 0$.

Let us consider the effective emissions in the transaction t

$$E_{eff,t}^S = \hat{E}_t^S\left\{1 - \left[1 - (1 + \frac{e_{t-1}^{lS}}{e_{t-1}^{uS}})\alpha\right]\frac{e_{t-1}^{uS}}{\hat{E}_t^{uS}}\right\} \tag{15}$$

We start with a simple observation.

Lemma 1. $E_{eff,t}^S = \hat{E}_t^S\left[1 - u^S(\alpha)\right]$

Proof

Taking into account that from (11)

$$\frac{e_t^{lS}}{e_t^{uS}} = \frac{d_t^{lS}}{d_t^{uS}} = \frac{d^{lS}}{d^{uS}}$$

and using the right hand side of (11) we have

$$E_{eff,t}^S = \hat{E}_t^S\left\{1 - \left[1 - (1 + \frac{d^{lS}}{d^{uS}})\alpha\right]R_0^{uS}\right\} = \hat{E}_t^S\left[1 - u^S(\alpha)\right]$$

which ends the proof. $\square$

We now pass to the bounds on amount of traded emissions. Let us stress that the bounds below give only a preferable amount of emissions to be bought in order to achieve a compliance with the risk α. If not enough emissions is bought in the transaction, the remaining emissions may be possibly bought in subsequent transactions. But we do not consider here situations where the buyer purchases more emissions than he needs.

Theorem 1. *A reasonable maximal amount of the traded estimated emissions in a transaction* $t \geq 1$ *between a seller S and a buyer B is given by*

$$\hat{E}_t^S \leq \min\left\{ \frac{\hat{l}_{t-1}^B - L^B}{1 - u^S(\alpha)}, \frac{L^S - \hat{l}_{t-1}^S}{1 + u^S(\alpha)} \right\}$$

and of the traded effective emissions by

$$E_{eff,t}^S \leq \min\left\{ \hat{l}_{t-1}^B - L^B, \frac{1 - u^S(\alpha)}{1 + u^S(\alpha)}(L^S - \hat{l}_{t-1}^S) \right\}$$

Proof

It is easy to calculate how many units of permits $\hat{E}_t^S$ the buyer should buy to become compliant with the risk α. Let us denote the recalculated expanded estimated emission of the buyer after the transaction $t - 1$ as

$$\hat{l}_{t-1}^B = \hat{x}_{t-1}^B + \left[1 - \alpha\left(1 + \frac{d_{t-1}^{lB}}{d_{t-1}^{uB}}\right)\right]d_{t-1}^{uB} = \hat{x}_{t-1}^B + d_{t-1}^{uB} - \alpha(d_{t-1}^{uB} + d_{t-1}^{lB})$$

To achieve the limit, the buyer has to satisfy after the transaction t the condition

$$\hat{x}_{t-1}^B - \hat{E}_t^S + \left[1 - \alpha\left(1 + \frac{d_{t-1}^{lB} + e_t^{lS}}{d_{t-1}^{uB} + e_t^{uS}}\right)\right](d_{t-1}^{uB} + e_t^{uS}) = L^B$$

where L^B is the buyer's limit. After simple algebraic manipulations the above equality can be transformed to

$$\hat{x}_{t-1}^B + d_{t-1}^{uB} - \alpha(d_{t-1}^{uB} + d_{t-1}^{lB}) - \left[\hat{E}_t^S - e_t^{uS} + \alpha(e_t^{uS} + e_t^{lS})\right] = L^B$$

The first part on the left hand equals $\hat{l}_{t-1}^B$ and the second, from Lemma 1, equals $E_{eff,t}^S$. Therefore, we obtain

$$\hat{l}_{t-1}^B - E_{eff,t}^S = L^B \tag{16}$$

Thus, the necessary amount of bought permits to achieve the compliance with the risk α is

$$\hat{E}_t^S = \frac{\hat{l}_{t-1}^B - L^B}{1 - u^S(\alpha)}$$

It is not optimal for the buyer to purchase more emissions than it is necessary. Thus, he will rather buy at most the amount equal to the right hand side

$$\hat{E}_t^S \leq \frac{\hat{l}_{t-1}^B - L^B}{1 - u^S(\alpha)} \tag{17}$$

The buyer may, however, prefer to buy, and certainly to pay, for the effective emissions. They have to satisfy the following simple condition to achieve the buyer's limit

$$E_{eff,t}^S \leq \hat{l}_{t-1}^B - L^B \tag{18}$$

The bounds (17) and (18) are the left hand values under the min operators in the theorem.

Considering now the seller, let us denote

$$\hat{l}_{t-1}^S = \hat{x}_{t-1}^S + \left[1 - \alpha\left(1 + \frac{d_{t-1}^{lS}}{d_{t-1}^{uS}}\right)\right] d_{t-1}^{uS}$$

After selling the estimated emissions, the seller should not exceed his limit, that is, it should hold

$$\hat{x}^S + \hat{E}_t^S + \left[1 - \alpha\left(1 + \frac{d_{t-1}^{lS} + e_t^{lS}}{d_{t-1}^{uS} + e_t^{uS}}\right)\right](d_{t-1}^{uS} + e_t^{uS}) \leq L^S$$

which can be transformed to

$$\hat{l}_{t-1}^S + \hat{E}_t^S + e_t^{uS} - \alpha(e_t^{uS} + e_t^{lS}) \leq L^S$$

Taking into account definition (15) and Lemma 1, we can write

$$\hat{E}_t^S + e_t^{uS} - \alpha(e_t^{uS} + e_t^{lS}) = \hat{E}_t^S\left[1 + u^S(\alpha)\right]$$

from where we get the condition on the maximum amount of estimated emissions to be sold

$$\hat{E}_t^S \leq \frac{L^S - \hat{l}_{t-1}^S}{1 + u^S(\alpha)} \tag{19}$$

Taking again into account Lemma 1, the condition (19) can be formulated in terms of the effective emissions as

$$E_{eff,t}^S \leq \frac{1 - u^S(\alpha)}{1 + u^S(\alpha)}(L^S - \hat{l}_{t-1}^S) \tag{20}$$

This bound is more important than the bound of the buyer, in the sense that if it is not satisfied, the seller looses the compliance with the risk α.

Now, combining (17) and (19), and then (18) and (20) we get the theorem thesis.   □

6   Market in Effective Emissions

Two kinds of emissions exist in the market outlined in the previous section, i.e. the estimated emissions and effective emissions. They have to be recalculated to each other during the trade. That is, both kinds of emissions exists at the market. Here we propose a market with only one kind of emissions, which are the effective emissions. In this market, only parties recalculate the estimated emissions to the effective ones and vice versa, for their internal purposes. In this market, the corrected limits of Definition 3 are used.

Then, let us consider a market in effective emissions, acting according to the following principles.

- When trading, the effective emissions and corrected limits are used.
- After a transaction, the seller adjusts his accumulated estimated emission according to the rule

$$\hat{x}_t^S = \hat{x}_{t-1}^S + \frac{E_{eff,t}^S}{1 - u^S(\alpha)} \tag{21}$$

- After a transaction, the buyer adjusts his accumulated estimated emission according to the rule

$$\hat{x}_t^B = \hat{x}_{t-1}^B - \frac{E_{eff,t}^S}{1 + u^B(\alpha)} \tag{22}$$

These definitions allow us to formulate simple bounds for a reasonable amount of effective emissions to be traded in a transaction.

Theorem 2. *A reasonable maximal amount of the effective emissions to be traded in a transaction t between a seller S and a buyer B is*

$$E_{eff,t}^S \leq \min\left\{ (1 + u^B(\alpha))(\hat{x}_{t-1}^B - \tilde{L}^B), (1 - u^S(\alpha))(\tilde{L}^S - \hat{x}_{t-1}^S) \right\} \tag{23}$$

Proof

Multiplying and dividing the right hand side of (18) by $1 + u^B(\alpha)$ we get

$$E_{eff,t}^S \leq \frac{1 + u^B(\alpha)}{1 + u^B(\alpha)}(\hat{l}_{t-1}^B - L^B)$$

Then, (23) follows from Definition (3) and Theorem 1.   □

Then we derive a basic property of the market.

Theorem 3. *A party is compliant with the risk α after transaction t, if and only if its accumulated estimated emission is not greater than its corrected limit*

$$\hat{x}_t \leq \tilde{L}$$

Proof

Let us first consider the seller. From (21), we have

$$\hat{x}_t^S = \hat{x}_0^S + \frac{\sum_{i=1}^t E_{eff,i}^S}{1 - u^S(\alpha)}$$

Then, from (15) we get

$$\hat{x}_t^S = \hat{x}_0^S + \sum_{i=1}^t \hat{E}_i^S$$

So, it is just the sum of the initial emission and all sold permits. All of them are associated with the seller's uncertainty. Thus

$$\hat{l}_t^S = \hat{x}_t^S[1 + u^S(\alpha)]$$

As the seller is compliant with the risk α after transaction t, that he satisfies the appropriate inequality similar to (7), then, using additionally Definition 3

$$\hat{x}_t^S \leq \frac{L^S}{1 + u^S(\alpha)} = \tilde{L}^S$$

This proves the "if" part of the theorem for the seller. To prove the "only if" part let us notice that the reasoning can be easily reversed. So, the theorem is true for the seller.

Let us now consider the buyer. Similarly as above, we have

$$\hat{x}_t^B = \hat{x}_0^B - \frac{\sum_{i=1}^t E_{eff,i}^{S_i}}{1 + u^B(\alpha)} = \frac{[1 + u^B(\alpha)]\hat{x}_0^B - \sum_{i=1}^t [1 - u^{S_i}(\alpha)]\hat{E}_i^{S_i}}{1 + u^B(\alpha)}$$

Now, considering iteratively transaction after transaction, it can be noticed that the numerator on the right hand side above, is equal to the left hand side of (9). Also this reasoning can be reversed. Thus

$$\hat{x}_t^B \leq \frac{L^B}{1 + u^B(\alpha)} = \tilde{L}^B$$

So, the theorem is also true for the buyer.

In the general case, we can order the buying transactions as the first $K < t$ transactions, without loosing generality. Then, considering only the first K transactions we know that the theorem is true. Treating now the estimated emissions and uncertainty spreads after first K transactions as a new starting point, and considering then the selling transactions we conclude from the former part of the proof that the theorem is true. This completes the proof of the theorem.   □

In conclusion, the organization of the market is as follows.

1. Before starting, the limits are recalculated to the corrected limits $\tilde{L}$ according to Definition 3.
2. In the trade, the parties negotiate the trading conditions taking into account the effective emissions E_{eff}. The amount of effective emission possessed by a party for selling is calculated as $E_{eff} = [1 - u^S(\alpha)](\tilde{L}^S - \hat{x}_{t-1}^S)$. The amount of effective emissions needed by the seller is calculated as $E_{eff} = [1 + u^B(\alpha)](\hat{x}_{t-1}^B - \tilde{L}^B)$.
3. After terminating the transactions both the seller and the buyer adjust their corrected estimated emissions according to (21) and (22), respectively.
4. To check the compliance, the present corrected estimated emissions are compared with the corrected limits.

In simulation of the trade, the reduction cost curves are expressed in terms of the accumulated estimated emissions. However, these prices have to be recalculated to the prices of the efficient emissions that can be compared with prices proposed by the trading partner. For the seller, we have $E_{eff} = [1 - u^S(\alpha)]\hat{x}^S$. Thus, one unit of the estimated emissions $\hat{x}^S$ is equivalent to $1 - u^S(\alpha)$ units of the effective emissions $E_{eff,t}$. Therefore, the following holds.

5. The price of one unit of efficient emissions c_{eff} and one unit of estimated emissions c^S for the seller are related as follow

$$c_{eff}[1 - u^S(\alpha)] = c^S$$

6. The price of one unit of efficient emissions c_{eff} and one unit of estimated emissions c^B for the buyer are related as follow

$$c_{eff}[1 + u^B(\alpha)] = c^B$$

As a simple corollary we note

$$\frac{c^S}{1 - u^S(\alpha)} = \frac{c^B}{1 + u^B(\alpha)}$$

7 Conclusions

The chapter deals with the problem of trading of pollutant emissions in the case when the observed emission values are highly uncertain with asymmetric uncertainty distributions. Asymmetric uncertainty of national greenhouse gases inventories is evidenced by recent investigations, and particularly by Monte Carlo simulations of uncertainty distributions.

In the market proposed in the chapter the limits are converted to so-called corrected limits, which are smaller than original limits. This is because of taking into consideration the unreported emissions due to uncertainty. Using the corrected limits, organization of the market in effective emissions is presented and its basic properties are proved. The market operates almost the same way as a usual market. The difference is that after each transaction the effective emissions have to be appropriately converted by the seller and the buyer in order to adjust the accumulated estimated emissions of both trading parties.

Acknowledgement. Comments by Dr Piotr Pałka and Dr Jarosław Stańczak helped us to improve an earlier version of the paper and are gratefully acknowledged.

References

1. Bartoszczuk, P., Horabik, J.: Tradable permit systems: Considering uncertainty in emission estimates. Water Air Soil Poll: Focus 7, 573–579 (2007)
2. Ermolieva, T., Ermoliev, Y., Fischer, G., Jonas, M., Makowski, M., Wagner, F.: Carbon emission trading and carbon taxes under uncertainty. Climatic Change 103, 277–289 (2010)
3. Gillenwater, M., Sussman, F., Cohen, J.: Practical policy applications of uncertainty analysis for national greenhouse gas inventories. Water Air Soil Poll: Focus 7, 451–474 (2007)
4. Godal, O., Ermolev, Y., Klaassen, G., Obersteiner, M.: Carbon trading with imperfectly observable emissions. Environ. Resour. Econ. 25, 151–169 (2003)
5. Graves, S.B., Ringuest, J.L.: Probabilistic dominance criteria for comparing uncertain alternatives: A tutorial. Omega 37, 346–357 (2009)
6. Hurteaux, M.D., Hungate, B.A., Koch, G.W.: Accounting for risk in valuing forest carbon offset. Carbon Balance and Management 4, 1 (2009),
   http://www.cbmjournal.com/content/4/1/1
7. Jonas, M., Gusti, M., Jęda, W., Nahorski, Z., Nilsson, S.: Comparison of preparatory signal detection techniques for consideration in the (post-)Kyoto policy process. In: Proc. 2nd Int. Wokshop Uncertainty in Greenhouse Gas Inventories, IIASA, Laxenburg, pp. 107–134 (2007)
8. Jonas, M., Marland, G., Winiwarter, W., White, T., Nahorski, Z., Bun, R., Nilsson, S.: Benefits of dealing with uncertainty in greenhouse gas inventories. Climatic Change 103, 3–18 (2010)
9. Jonas, M., Nilsson, S.: Prior to economic treatment of emissions and their uncertainties under the Kyoto Protocol: Scientific uncertainties that must be kept in mind. Water Air Soil Poll.: Focus 7, 495–511 (2007)
10. Lieberman, D., Jonas, M., Nahorski, Z., Nilsson, S. (eds.): Accounting for Climate Change. Uncertainty in Greenhouse Gas Inventories – Verification, Compliance, and Trading. Springer, Dordrecht (2007)
11. Mignone, B.K., Hurteau, M.D., Chen, Y., Sohngen, B.: Carbon offsets, reversal risk and US climate policy. Carbon Balance and Management 4, 3 (2009),
    http://www.cbmjournal.com/content/4/1/3
12. Monni, S., Syri, S., Pipatti, R., Savolainen, I.: Extension of EU emissions trading scheme to other sectors and gases: consequences for uncertainty of total tradable amount. Water Air Soil Poll.: Focus 7, 529–538 (2007)
13. Nahorski, Z., Jęda, W., Jonas, M.: Coping with uncertainty in verification of the Kyoto obligations. In: Studziński, J., Drelichowski, L., Hryniewicz, O. (red.) Zastosowanie informatyki i analizy systemowej w zarządzaniu, IBS PAN, Warszawa, pp. 305–317 (2003)
14. Nahorski, Z., Horabik, J., Jonas, M.: Compliance and emission trading under the Kyoto Protocol: Rules for uncertain inventories. Water Air Soil Poll.: Focus 7, 539–558 (2007)

15. Nahorski, Z., Horabik, J.: Compliance and emission trading rules for uncertain estimates of inventory uncertainty. In: Proc. 2nd Int. Wokshop Uncertainty in Greenhouse Gas Inventories, IIASA, Laxenburg, pp. 149–161 (2007)
16. Nahorski, Z., Horabik, J.: Greenhouse gas emission permit trading with different uncertainties in emission sources. J. Energ. Eng.-ASCE 134, 47–52 (2008)
17. Nahorski, Z., Horabik, J.: Compliance and emission trading rules for asymmetric emission uncertainty estimates. Climatic Change 103, 303–325 (2010)
18. Nahorski, Z., Stańczak, J., Pałka, P.: Application of multi-commodity market model for greenhouse gases emission permit trading. In: Kaleta, M., Traczyk, T. (eds.) Modelling Multi-Commodity Trade: Information Exchange Methods. Warsaw University of Technology, Institute of Control and Computation Engineering, Warsaw, pp. 108–119 (2010)
19. Ramirez, A.R., de Keizer, C., van der Sluijs, J.P.: Monte Carlo analysis of uncertainties in the Netherlands greenhouse gas emission inventory for 1990-2004. Report NWS-E-2006-58. Copernicus Institute for Sustainable Development and Innovation. Utrecht (2006), http://www.chem.uu.nl/nws/www/publica/publicaties2006/E2006-58.pdf
20. Roy, A.D.: Safety first and the holding of assets. Econometrica 20, 431–449 (1952)
21. Stańczak, J., Bartoszczuk, P.: CO_2 emission trading model with trading prices. Climatic Change 103, 291–301 (2010)
22. Winiwater, W.: National greenhouse gas inventories: understanding uncertainties versus potential for improving reliability. Water Air Soil Poll.: Focus 7, 443–450 (2004)
23. Winiwarter, W., Muik, B.: Statistical dependences in input data of national GHG emission inventories: effects on the overall GHG uncertainty and related policy issues. Presentation at 2nd Int. Workshop Uncertainty in Greenhouse Gas Inventories, September 27-28. IIASA, Laxenburg (2007)

Application of Multi-commodity Market Model for Greenhouse Gases Emission Permits Trading

Zbigniew Nahorski, Jarosław Stańczak, and Piotr Pałka

Abstract. Greenhouse gases emission permits trading can be modeled using the multi-agent platform for multi-commodity exchange. A simulation of this kind of trade is described in the paper. A party can use one of two strategies to find a good partner to achieve best gain: (i) bilateral trade with a randomly chosen feasible partner, (ii) a tender. In the tender trade, parties submit offers to the current tender operator; the tender operator chooses the offer of the party that maximizes his gain. The results of simulation are presented.

1 Introduction

The market behavior is usually analyzed either using a static optimization model or a game-theoretic approach. In both cases it is assumed that an appropriate information is available to the parties acting on the market. Recently, agent-based models are often used to investigate the dynamics of the market behavior using a simulation approach. The parties are represented by intelligent programming agents who negotiate and trade the goods according to given market rules and the information available to the agents. This way, the method presented in this chapter need not assume an ideal market. Neither the equilibrium prices are assumed in the trading, nor the dynamic trading without taking prices into account, like in [4, 5], are considered. A more sophisticated, dynamic market model is introduced, with possibility

Zbigniew Nahorski
Systems Research Institute, Polish Academy of Sciences
e-mail: Zbigniew.Nahorski@ibspan.waw.pl

Jarosław Stańczak
Polish Academy of Sciences, System Research Institute
e-mail: Jaroslaw.Stanczak@ibspan.waw.pl

Piotr Pałka
Warsaw University of Technology, Institute of Control and Computation Engineering
e-mail: P.Palka@ia.pw.edu.pl

M. Kaleta & T. Traczyk (Eds.): Modeling Multi-commodity Trade, AISC 121, pp. 165–177.
springerlink.com   © Springer-Verlag Berlin Heidelberg 2012

of price negotiation and the influence of real prices on the agent behavior (similar assumptions can be found in [2]). The number of transactions between the start of the market and equilibrium is not known in advance. Only transactions profitable for both participants are accepted during simulations. Each transaction that is profitable for both parties moves the market toward equilibrium. Transaction can be simulated using a multi-agent platform for multi-commodity exchange [7]. This gives a possibility to trace activities of agents participating in the market and to build more realistic model of conducted transactions. In particular, intelligent agents are considered and trading strategies of the agents are analyzed and optimized. Each agent minimizes its own objective function, which is the costs including spendings on emission reduction plus expenditures for the permits. This approach takes into account the purchase/sale price of permits, which considerably influences the profitability of transactions and the decision to buy/sell permits, i.e. whether it is better to reduce emissions or to buy permits. Obtained so far results of simulations are close to the equilibrium point calculated for the problem where the common costs for all trading agents are minimized using optimization methods.

2 Centralized Model

Let us consider a market with N parties trading the emission permits. Each party has been pre-allocated K_n permits, $n = 1, \ldots, N$, usually called "the Kyoto targets". At the compliance time the party must not emit more than the number of permits it possesses. However, it may freely sell or buy permits to achieve the target. We denote by x_n the emission of the nth party, by x_n^0 the emissions of the nth party, if no trade is performed (BAU – business as usual emissions), and by y_n the traded permits. Emission are nonnegative, $x_n \geq 0$, but number of permits y_n may be positive, when bought, or negative, when sold. To achieve reduction, the following condition before starting the trade has to be satisfied

$$\sum_{n=1}^{N} x_n^0 > \sum_{n=1}^{N} K_n$$

When designing the market, the central planner perceives the use of emission trading system as minimization of a social cost function

$$F(\mathbf{x}) = \sum_{n=1}^{N} c_n(x_n), \quad \mathbf{x} = (x_1, \ldots, x_N) \tag{1}$$

when $c_n(x_n), c_n'(x_n) < 0$, is the cost born by the nth party to reduce the emission to the level x_n [1], and subject to

$$\sum_{n=1}^{N} x_n = K_0$$

---

[1] Notice that $c_n(x_n^0) = 0$, as no reduction costs are necessary for the BAU emission.

where K_0 is the total allowable emission of market participants. Let us notice that equality above is the consequence of an observation that it is not optimal to have emissions strictly less than K_0.

Using the Lagrange method, the necessary condition of optimality is

$$\lambda = -c'_n(x_n^{opt}), \quad n = 1,\ldots,N; \qquad \sum_{n=1}^{N} x_n^{opt} = K_0 \tag{2}$$

Thus, at the optimal emissions the marginal costs of all parties are equal and λ is an optimal price.

We see that for known functions $c_n(x_n)$ for all n, the optimal price λ can be found. However, the functions $c_n(x_n)$ are not known to the central planner. Thus, the problem is left to be solved by the market, where each party is looking to minimize its cost of reducing the emission and buying/selling the permits y_n. That is, the nth party minimizes the function

$$f_n(x_n, y_n) = c_n(x_n) + \lambda_t y_n \tag{3}$$

subject to

$$x_n \leq K_n + y_n$$

Above, $\lambda_t \geq 0$ is the price of one unit of permits in the tth transaction. The prices in consequent transactions may vary, as the optimal price is not known to the trading parties, and it can only be possibly estimated, if the market actual prices eventually converge. As the trade is continuing in time, some parties negotiate the transactions earlier and some later, not knowing the optimal price λ. Just, before the prices converge, the parties have to cope with the uncertainty in earning/loosing money in the trade.

3   Dynamic Bilateral Trading

In [4], an approach to simulate the bilateral exchange of permits between parties has been proposed, called the dynamic bilateral trading. The idea is that two parties meet randomly and sell/buy permits, if it is feasible for both parties. This happens when the marginal costs of the parties differ, i.e. the first condition in (2) is not satisfied for all parties. It is a simple observation that each such transaction makes the cost function (1) smaller, see [5]. As the function (1) is constrained from below by 0, and the sequence of the values of the function (1) is decreasing with each transaction, then it converges, although not necessarily to the global minimum of (1). In the original paper [3] it is assumed that the number of exchanged permits goes to zero (albeit not too quickly) and then it is proved that the sequence converges to the global minimum with probability 1. Thus, the approach actually uses a stochastic optimization method. That approach does not take into account the prices of the permits. Neither it is considered, if the parties really need to buy or sell permits.

A modification of this problem, in which the prices are allowed for, has been presented in [9]. As before, only feasible transactions are considered. But the price π of the transaction is drawn at random from the interval constrained by the marginal costs of the trading parties. The number of trading permits is drawn randomly as well. It is considered that the parties look during the trading at the financial result in each transaction, viz. the difference between the gain from selling the permits and costs of reducing the emissions. This can be written as

$$g_n^t = y_n^t \pi_n^t - \left[c_n(x_n^{t-1} + y_n^t) - c_n(x_n^{t-1}) \right]$$

A party can choose one of two strategies to find a good partner for winning the above competition. One is to choose a partner randomly, like in [4], if the trade is profitable for both partners. The second strategy is a tender. In the latter case, a randomly drawn number of permits is offered for sale and this feasible partner is chosen, for whom the gain is the greatest, for a randomly drawn price. A strategy is drawn randomly with probability assigned each time by an evolutionary algorithm, as described in more details in the sequel. The iteration number t is incremented each time a transaction between parties is completed.

Let us also notice that the algorithm does not start from the emissions x_n^s but K_n. This is connected with the fact that actually the algorithm does not simulate the real trade but maximizes iteratively the common function (1) and is, up to some details, a subversion of a more general algorithm from [4], where the costs of optimal solution for each party is calculated using the equilibrium price λ. In [9], the costs are sums of incomes from all transactions conducted by the party.

In this chapter we follow a general idea of the above algorithm. However, only local goal functions are considered and more advanced methods of negotiating prices are applied. Each party, also called an agent to conform with the multi-agent systems nomenclature, maximizes only his own income g_n^t.

A party initiating a consecutive transaction is drawn randomly. The number of traded permits is drawn randomly, as before. However, in this chapter, in the bilateral trade the price of a possible transaction is negotiated with the prospective partner. There are several negotiation strategies to be chosen by a party and the probability of the choice evolves according to the evolutionary algorithm. One of the simpler is the linear bidding when each negotiating party steps off a constant sum from its price in consequent biddings until both parties meet.

The formal definition of the market is as follows. The randomly drawn nth party enters the trade and randomly draws y_n^t emission permits for trade, selling or buying, as appropriate. Then it tries to maximize its goal function by negotiating transactions with chosen randomly parties, among $m = 1, \ldots, N$, $m \neq n$, while keeping the bounds

$$x_n^t = x_n^{t-1} + y_n^t, \ x_n^0 = K_n \tag{4}$$

$$x_n^t \leq K_n + \sum_{\tau=1}^{t} y_n^\tau, \ n = 1, \ldots, N \tag{5}$$

$$t = 1, \ldots, T \tag{6}$$

A limitation on the number of traded permits is also introduced. The market simulation (transactions between parties) stops after T transactions due to lack of possible profits to be gained.

4   Application of M^3 Model to GHG Emission Permits Market

The application of M^3 model was performed using Java implementation of multi-agent platform for multi-commodity exchange described in chapter *Application of the multi-agent systems in the context of the multi-commodity market model M^3*. We describe only modifications, which had to be done to get M^3 platform application working properly, according to description.

4.1   Agents Structure

Three agent roles were defined during simulation of GHG emission permits market: the Morris Column role, negotiator role (with two initially set behaviors: activeness and passiveness) and the auction operator role (for the tender trading). The negotiator and the auction operator roles were applied to specific agent dynamically, so one agent has various roles, depending on situation.

Each agent represents single party, which is guided by its own interests. The individual party comes to interact with others, motivated by the wish to achieve certain gains from the exchange of permits, i.e. to reduce the costs including trading plus expenditures for the permits. The overarching goals of the system can be identified with the objectives of the central planner, i.e. to minimize the social cost – the sum of costs to reduce the emission limits. However, we must remember that these goals need not always be met.

4.2   Trading Methods: Bilateral Contracts and Tender

Each trading method needs different agent role specifications, and also different behaviors.

For the bilateral contract trading, we assume that an agent can choose to be a passive negotiator, i.e. one who submits its willingness to negotiate to the Morris Column and waits for negotiation partners; or an agent can choose to be an active negotiator, i.e. one who searches for negotiating partners by querying the Morris Column. We assume, that agents choose to be passive or active randomly. Agents perform negotiations in randomly established pairs, and conducts bilateral contracts depending on expected profits.

In the case of tender, we assume that one agent (in a specified moment), plays a role of an auction operator, and the rest of agents who entered the auction, plays a bidder roles. Roles are assigned dynamically and change during the trading. The selection of the auction operator among a number of agents in the distributed environment is a problem. A solution can be to apply the bully election algorithm [8] for selection of the operator from a group of agents. Thus, the operator is chosen randomly, the rest of agents submits bids for selling or buying a number of permits, at the price specified by the bidder. The operator chooses the most profitable bid, taking into account his preferences. Afterwards, a new operator is chosen randomly and the process is repeated again.

4.3   Agent Behaviors

The agent behavior is represented by a predetermined sequence of steps, which lead to achievement of a particular goal. JADE multi-agent environment (which is implementation of FIPA standards – see Chapter *Application of the multi-agent systems in the context of the multi-commodity market model M^3*) provides a class (named jade.Behavior) to implement specific agents behaviors. In the GHG emission permits market, we implemented behaviors corresponding to the action of agents both for the bilateral negotiations and for the tender.

4.4   Communicative Acts and Message Content Details

Application of M^3 platform to the GHG emission trading market is restricted to: (i) logging negotiation proposals sent to parties in the course of negotiations, (ii) logging the offers sent to party, which is currently chosen the operator of the tender. Moreover, the operator has to submit its current trading proposal to the Morris Column agent.

4.4.1   Operator's Trading Proposal

The party, which is the current tender operator sends a CFP message to the Morris Column agent. The element `Announce` is included in the content of this message (see Listing 1). Party, which is willing to submit an offer to the tender operator, have to place in the message the `in-reply-to 2-74` parameter.

Listing 1 Exemplary *cfp* message send by tender operator

```
(cfp
 :sender USA@power-dell:1099/JADE
 :receiver MorrisColumn@power-dell:1099/JADE
 :content (
    <Announce agentId="USA.power-dell.1099.JADE"
          xmlns:m3="http://www.openM3.org/m3">
      <anMarket offerType="ask" type="tender"
             id="AID-1354116916">
        <m3:tradedCommodity ref="CO2-emission-permit"/>
        <m3:tradingPeriod ref="Year2010" />
      </anMarket>
    </Announce>
 :language M3XML
 :ontology Greenhouse Gases
 :reply-to 2-74 )
```

4.4.2   Parties Willingness to Negotiate

The negotiating party, which plays a passive role, should submit its willingness for negotiating (see Listing 2). This is done by sending the CFP message to the Morris Column agent. A party, which is willing to take part in negotiation, should place in the message the `in-reply-to 5-37` parameter.

Listing 2 Exemplary *cfp* message send by passive negotiator

```
(cfp
 :sender EU@power-dell:1099/JADE
 :receiver MorrisColumn@power-dell:1099/JADE
 :content (
    <Announce agentId="EU.power-dell.1099.JADE"
          xmlns:m3="http://www.openM3.org/m3">
      <anMarket offerType="bid" type="Negotiations"
                id="AID-1354116916">
        <m3:tradedCommodity ref="CO2-emission-permit"/>
        <m3:tradingPeriod ref="Year2010" />
      </anMarket>
    </Announce>
 :language M3XML
 :ontology Greenhouse Gases
 :reply-to 5-37 )
```

4.4.3   Negotiation Proposal

We assume that a negotiation proposal is sent to a negotiation partner in the communicative act *propose*. The message, presented in Listing 3 should be read as follows: party `EEFSU@power-dell:1099/JADE` sends to negotiation partner `Japan@power-dell:1099/JADE` a negotiation proposal for selling up to 4.2 units of `CO2 emission permit` at the unit price 77.32. A single party may negotiate with multiple partners, so to identify a particular negotiation process, the suitable identifier is needed. We use the `conversation id` for this purpose.

Listing 3 Exemplary *propose* message send by negotiating agent

```
(propose
 :sender EEFSU@power-dell:1099/JADE
 :receiver Japan@power-dell:1099/JADE
 :content (
     <m3:Offers xmlns:m3="http://www.openM3.org/m3">
       <m3:Offer id="OID732819423" offeredPrice="77.32">
         <m3:name>Negotiation Offer No.1</m3:name>
         <m3:volumeRange minValue="0" maxValue="4.2"/>
         <m3:offeredBy ref="EEFSU.power-dell.1099.JADE"/>
         <m3:ElementaryOffer>
           <m3:offeredCommodity ref="CO2-emission-permit"
                                shareFactor="1.0"/>
         </m3:ElementaryOffer>
       </m3:Offer>
     </m3:Offers> )
 :language M3XML
 :ontology Greenhouse Gases
 :conversation-id 3-86 )
```

Let us assume, for example, that `Japan@power-dell:1099/JADE` is willing to accept this proposal. It sends the message presented in Listing 4. Note, that the `conversation-id` parameter is the same as in the previous message. Moreover, we can see that in the message body the attribute `acceptedVolume` appears, which is the amount of the approved commodity.

Listing 4 Exemplary *accept-proposal* message send by negotiating agent

```
(accept-proposal
 :sender Japan@power-dell:1099/JADE
 :receiver EEFSU@power-dell:1099/JADE
 :content (
     <m3:Offers xmlns:m3="http://www.openM3.org/m3">
       <m3:Offer id="OID732819423" offeredPrice="77.32"
             acceptedVolume="4.2" >
         <m3:name>Negotiation Offer No.2</m3:name>
         <m3:volumeRange minValue="0" maxValue="4.2"/>
         <m3:offeredBy ref="Japan.power-dell.1099.JADE"/>
         <m3:ElementaryOffer>
           <m3:offeredCommodity ref="CO2-emission-permit"
               shareFactor="-1.0"/>
         </m3:ElementaryOffer>
       </m3:Offer>
     </m3:Offers> )
 :language M3XML
 :ontology Greenhouse Gases
 :conversation-id 3-86 )
```

4.4.4 Offer

A party which is willing to submit an offer to the current tender operator, should send him the *propose* communicative act. Note that this is the same communicative act as in the bilateral negotiation case. The receiver must distinguish between the meaning of these acts on the basis of its current role, and on the parameters `in-reply-to` and `conversation-id`.

4.5 Parties Strategic Behavior

Parties' (or agents') intelligence is inextricably related to the fact that each participant is trying to achieve the highest possible profits. Thus, an analysis, which neglect the possibility of parties' strategic behavior, is not sufficient.

More sophisticated strategies combined with adequate conditions may lead a party to reach higher profits, comparing with application of a simple strategy of just reporting the fixed price by an agent. We say that such a party has a market power, and even more, that it abuses it. Parties who abuse of market power may harm other, weaker parties, by depriving them of the profit from the transaction. Moreover, the market efficiency decreases, and the weaker parties can abandon the market. So the analysis of influence of different strategies and trading rules in the emission permits market is an important issue. In this chapter we present results of simulation, when particular agents are intelligent and autonomous. An agent is intelligent due to using appropriate strategies. The goal of using them is to allow the agent to obtain the highest profits. Several simple strategies will be presented and the results of their use will be analyzed.

Agent's behavior described earlier does not underline the efforts of agents to achieve higher profits. For that, we implemented three simple strategies. The first one – the frank behavior, assumes that a party does not behave strategically at all. Two remaining strategies assumes more sophisticated behavior of parties. Both of them are based on establishing a factor, which allows the agent to decide whether to complete the transaction at the present stage of negotiation, or to present another, more competitive offer. All the transactions (including those rejected) are logged by the agents. When the agent prepares the offer, it checks, how many previous similar transactions were accepted, and how many were rejected. On this basis it determines the factor, on the basis of which it decides how the price may differ from the marginal price. Although this strategy is very simple, it gives quite telling results. In the sequel of this chapter we will call this strategy the Successful Transaction Number (STN) Strategy. The second strategy is based on checking how many failures occurred since the last successful transaction. If this number is greater than a fixed parameter, then the factor, which determines how the price may differ from the marginal price, changes. In the sequel of this chapter we will call it the Failure Detection (FD) Strategy.

5 Results

The simulation was carried out on the case study described in the [6]. The case study assumes participation of five parties: USA, EU, Japan, CANZ (Canada, Australia, New Zealand) and EEFSU (East Europe and Former Soviet Union). These parties should comply with the Kyoto requirements, so most of them should reduce CO_2 emission, except of EFFSU, whose initial emission is smaller then the target. The parties have been specified an emission limit K_i, initial emission x_i^0, and the cost reduction function C_i (index i defines the party). As set forth in [6], the parties cost reduction functions are well approximated by a square function of the emission reduction size – see (7)

$$C_i = \begin{cases} a_i(x_i^0 - x_i)^2 & \text{for} \quad x_i < x_i^0 \\ 0 & \text{for} \quad x_i \geq x_i^0 \end{cases} \tag{7}$$

The marginal cost of a permit (the shadow price), is the derivative of the cost reduction function at the appropriate level of emission (see (8))

$$c_i = \begin{cases} 2a_i(x_i^0 - x_i) & \text{for} \quad x_i < x_i^0 \\ \min & \text{for} \quad x_i \geq x_i^0 \end{cases} \tag{8}$$

where x_i stands for the current emission. In the beginning of the simulation, we assume that each party starts from its Kyoto limit $x_i = K_i$. The *min* is the minimum price value, and it was introduced to get more realistic conditions in the price calculation. The data for the case study are presented in the Table 1.

Table 1 Data for the case study – parameters of function of cost reduction

Party	Current emission x_i^0 [MtC/y]	Cost function parameter a_i [MUSD/(MtC/y)^2]	Kyoto limit K_i [MtC/y]
USA	1 820.3	0.2755	1 251.0
EU	1 038.0	0.9065	860.0
Japan	350.0	2.4665	258.0
CANZ	312.7	1.1080	215.0
EEFSU	898.6	0.7845	1314.0

5.1   Results for "Frank" Behavior

Performed simulation of the greenhouse gases emission permits market are based on the multi-agent platform for the multi-commodity exchange (results are in Tables 2-3). As can be noticed, the results are very similar to the results obtained with the centralized market for the greenhouse gases permits [1].

Let us notice that for the tender trade, the sum of permit costs (for parties that bought the permits) is smaller than in the case of bilateral transaction. At the same time, the EEFSU party earns less from selling permits. It is due to the fact that in the tender trade conducted by buyers the contracts are more profitable for the buyers than in the bilateral transaction. In Fig. 1 we can see the trajectory of prices during one simulation of bilateral negotiations. As we can see, the prices converge to the equilibrium.

There are also several more factors, that can influence costs of reached contracts. The most important one is the strategy of setting price bids in negotiations. There are many possibilities to do it, some of them are described earlier in this chapter. This

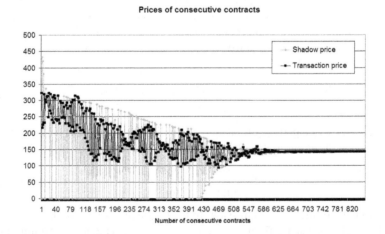

Fig. 1 Trajectory of consecutive transaction and shadow prices, during single simulation of bilateral negotiations

Table 2 Results of simulation using multi-agent system, bilateral contracts

Party	Final emission	Last transaction price	Final price (shadow price)	No. of traded permits	Cost of a permit	Reduction cost
	[MtC/y]	[USD/tC]	[USD/tC]	[MtC/y]	[MUSD/y]	[MUSD/y]
USA	1 561.5	142.6	142.6	310.5	52 778.4	18 446.2
EU	959.3	142.6	142.7	99.3	26 769.2	5 612.9
Japan	321.1	142.6	142.6	63.1	6 182.6	2 062.0
CANZ	248.4	142.6	142.6	33.4	10 164.5	4 586.9
EEFSU	807.7	142.6	142.6	-506.3	-95 894.8	6 482.1

Table 3 Results of simulation using multi-agent system, the tender

Party	Final emission	Last transaction price	Final price (shadow price)	No. of traded permits	Cost of a permit	Reduction cost
	[MtC/y]	[USD/tC]	[USD/tC]	[MtC/y]	[MUSD/y]	[MUSD/y]
USA	1 561.4	142.9	142.7	310.4	67 843.6	18 466.6
EU	959.4	142.7	142.5	99.4	21 316.1	5 600.3
Japan	321.1	142.1	142.6	63.1	15 904.3	2 060.0
CANZ	248.3	142.5	142.7	33.3	2 359.6	4 595.3
EEFSU	807.8	142.6	142.5	-506.2	-107 423.6	6 467.9

problem is particularly important for EEFSU. This party has the level of emission lower than the Kyoto target and its reduction costs are equal to 0 for the most of transactions. Of course, the real party does not want to sell permits for free. But it is difficult to set acceptable price level for transactions of this party – theoretically any positive price is profitable.

5.2 Results for Strategic Behavior

As stated before, two strategies were used during simulation of emission permits market. The Successful Transaction Number (STN) Strategy and the Failure Detection (FD) Strategy. In both strategies we assume that an agent can not loose when offering a price smaller or greater than the shadow price, nor that it can increase (or decrease in the case of the purchase) the shadow price by more than five times.

Application of both strategies resulted in the same equilibrium point as in the case where agents do not behave strategically (see Tables 2-3). The difference is in the number of conducted contracts and in overall cost values (see Fig. 2).

As we can see, strategies give different results. Application of each strategy gives various benefits for particular parties. For example, strategy FD is better for EU and CANZ parties, while strategy STN is better for EEFSU. Nevertheless, for USA, application of strategy depends strongly of the type of trade. For bilateral trading the better strategy is STN, for the tender the better strategy is FD.

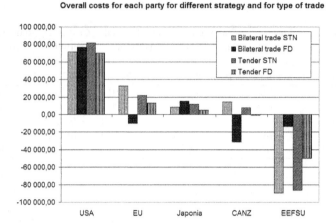

Fig. 2 Overall costs for each party for different strategies and for different types of trade

6   Conclusion

Results obtained in this chapter show, that application of M^3 and the multi-agent platform for multi-commodity trade for the problem of greenhouse gases emission permits trading is possible. The multi-agent platform seems to be a suitable tool for analyzing economic phenomena, especially for dynamic market models. It is also possible to integrate some elements of optimization or learning of trading strategies by independent agents, which is possible to do in multi-agent systems. Economic systems can be quite easily modeled, simulated and controlled using these kind of artificial intelligence tools.

The results obtained using the multi-agent trade are quite similar to the results obtained for the centralized market for the greenhouse gases permits, like presented in [6]. The values of final emissions, numbers of traded permits, final shadow prices and the reduction costs are almost identical.

References

1. Bartoszczuk, P., Horabik, J.: Tradable permit systems: Considering uncertainty in emission estimates. Water Air Soil Poll.: Focus 7, 573–579 (2007)
2. Bonatti, M., Ermoliev, Y., Gaivoronski, A.: Modeling of multi-agent market systems in the presence of uncertainty: The case of information economy. Robot. Auton. Syst. 3-4, 93–113 (1998)
3. Ermoliev, Y., Klaasen, G., Nentjes, A.: Adaptive cost-effective ambient charges under incomplete information. J. Environ. Econ. Manag. 31, 37–48 (1996)
4. Ermoliev, Y., Michalevich, M., Nentjes, A.: Markets for tradeable emission and ambient permits: A dynamic approach. Environ. Resour. Econ. 15, 39–56 (2000)

5. Ermolieva, T., Ermoliev, Y., Fischer, G., Jonas, M., Makowski, M.: Cost effective and environmentally safe emission trading under uncertainty. Lect. Notes Econ. Math., vol. 633, pp. 79–99 (2010)
6. Horabik, J.: On the costs of reducing ghg emissions and its underlying uncertainties in the context of carbon trading. Tech. Rep. IR-07-036, International Institute for Applied Systems Analysis, Laxenburg (2007)
7. Kaleta, M., Pałka, P., Toczyłowski, E., Traczyk, T.: Electronic Trading on Electricity Markets within a Multi-agent Framework. In: Nguyen, N.T., Kowalczyk, R., Chen, S.-M. (eds.) ICCCI 2009. LNCS, vol. 5796, pp. 788–799. Springer, Heidelberg (2009)
8. Mamun, Q.E.K., Masum, S.M., Mustafa, M.A.R.: Modified bully algorithm for electing coordinator in distributed systems. In: Proc. 3rd WSEAS International Conference on Software Engineering, Parallel and Distributed Systems (2004)
9. Stańczak, J., Bartoszczuk, P.: CO_2 emission trading model with trading prices. Climatic Change 103, 291–301 (2010)

Modelling Virtual Network Market Data with Open Multi-commodity Market Model

Janusz Granat, Kamil Kołtyś, and Krzysztof Pieńkosz

Abstract. In this chapter the Multi-commodity Market Model is applied for modelling virtual network market data. Two different auction models supporting VPN service trading are considered. The first one enables to specify VPN demand requirements by the means of pipe model, while the second one leverages the hose model to describe VPN service in the buy offer. The market data for both auctions can be easily implemented in M^3 using the bundled offer and network virtualization concepts.

1 Introduction

Advances in communication technology stimulate the development of new class of markets. One of such markets is the communication bandwidth market, or in general telecommunication resources market. A dominant form of trading on the market of the telecommunications networks resources are bilateral contracts. We can distinguish real-time resource trading or fixed-term contracts.

The usage of a communication network usually differs at different times of a day, because each user has different usage patterns. It causes variations in network load. The bandwidth market allows on demand provisioning of bandwidth, which leads to real-time markets. Such markets and systems that supports functioning of such

Janusz Granat
Warsaw University of Technology, Institute of Control and Computation Engineering
e-mail: J.Granat@ia.pw.edu.pl

Kamil Kołtyś
Warsaw University of Technology, Institute of Control and Computation Engineering
e-mail: K.J.Koltys@elka.pw.edu.pl

Krzysztof Pieńkosz
Warsaw University of Technology, Institute of Control and Computation Engineering
e-mail: K.Pienkosz@ia.pw.edu.pl

M. Kaleta & T. Traczyk (Eds.): Modeling Multi-commodity Trade, AISC 121, pp. 179–191.
springerlink.com

market are very complex and require continuous monitoring. Market surveillance plays important mechanism roles in constructing this category of market models.

The important trends in management of future networks are network virtualizations. This is an approach, where several network instances can co-exist on a common physical network infrastructure. Current technologies such as Virtual Private Networks (VPNs), provide only traffic isolation. There is new research area that focuses on development of a new approaches for network virtualization as well as new business models for network provisioning and operations. In new business models, there is a need for advanced methods for managing all aspects of the contracts.

In this chapter we will focus on virtual private networks. However, more advanced resources management patterns are also possible.

2 Virtual Private Networks

Virtual Private Network (VPN) is a telecommunication service established over a public network between geographically dispersed customer endpoints. The goal is to provide VPN users with service comparable to dedicated private lines. To realize VPN services, sufficient amount of bandwidth of shared network resources must be reserved to satisfy traffic demand pattern specified by customers.

The basic way of representing the set of traffic demands values of VPN is in the form of the *pipe* model. It requires that VPN customer specifies, for each pair of endpoints, the maximum demand volume. This approach is applicable to VPNs for which the customer knows in advance the exact traffic demands matrix. As the number of endpoints per VPN is constantly growing, the traffic demands patterns are becoming more and more complex. Therefore, for some VPNs, it is almost impossible to predict maximum value of each traffic demand. In [1] authors proposed the *hose* VPN model, in which VPN customer specifies aggregate requirements per VPN endpoint, and not for each pair of endpoints. In comparison with the pipe model, the hose model provides a simpler and more flexible way of VPN traffic demands specification. The customer only defines ingress and egress bandwidths of VPN endpoints, which can be more easily predicted than the traffic demands between each pair of endpoints. However, it is much more difficult to find efficient bandwidth reservation for the hose model. In the pipe model the demands between VPN endpoints are specified by customer, whereas in the hose model all traffic demand patterns satisfying ingress and egress requirements must be considered.

Currently, the bandwidth trading for VPN provisioning are in the form of bilateral agreements in which two participants negotiate the contract terms. The negotiations are complex, nontransparent and time consuming. The buyer that wants to purchase bandwidth between several nodes, connected by a set of links owned by different providers, must independently negotiate with all of them. If the negotiation with one of them fails, the buyer will get useless bandwidth as it will not ensure the connection between all selected nodes. Also, even if the buyer manages to purchase bandwidth that ensures connectivity between all VPN endpoints, there is a risk that VPN service could be provided by much more cheaper set of links. Thus there is a need of designing more sophisticated market mechanisms that will not have such severe drawbacks that are involved with bilateral agreements.

In [6] the convenient auction mechanism for trading VPNs in the pipe model is proposed. It assumes that the sell offers regard to single telecommunication links and buy offers to VPNs. Each sell offer specifies the maximum volume of bandwidth offered to sell at particular link and minimum unit price of this bandwidth. Each buy offer contains the maximum price that buyer is willing to pay for VPN and the specification of VPN traffic demands between every pair of its endpoints. It is assumed that sell and buy offers can be partially accepted. The auction mechanism matches sell and buy offers aiming at maximization of social surplus. The optimal allocation is efficiently determined by solving linear programming (LP) model. The essential feature of the mechanism is that the customer whose offer is accepted purchases the bandwidth for the complete set of links required for realizing VPN demands. If the buy offer is partially accepted the set of links is still complete, but the allocated amount of bandwidth at these links is proportionally reduced. This auction mechanism will be further called VPN_PIPE auction model.

The auction mechanisms for VPNs specified in the hose model are much more complex. The buy offers contain, in this case, the maximum price for VPN and the required ingress and egress bandwidth of every endpoint. Auction model for trading VPNs with the hose specification is presented in [7, 8]. Similarly to VPN_PIPE auction model, this mechanism determines the allocation that maximizes the social surplus. The sell and buy offers can be partially accepted but mechanism guarantees that even if buy offer is partially accepted the buyer will be able to realize all traffic demands specified in the hose model with proportionally reduced egress and ingress bandwidth of each VPN endpoint. The auction mechanism can be formulated as LP optimization problem and can be efficiently solved by applying column generation technique. Further this auction mechanism will be referred to as VPN_HOSE auction model.

3   Implementing VPN Auction Models in M^3

In this section we show how to implement the data for VPN_PIPE and VPN_HOSE auction models in M^3 data model. We will benefit from the work presented in [4] where the application of M^3 for representing data of model for balancing communication bandwidth trade (BCBT) [10] is presented. Authors show how main elements of M^3 data model such as infrastructure, time structure, market entities, commodities and offers, can be used for specifying all data for the BCBT model. Both VPN_PIPE and VPN_HOSE auction models are based on the BCBT model, in particular they require common subset of market data, such as: data relating physical network and data associated with the supply side. Also data concerning the results of VPN_PIPE and VPN_HOSE models can be represented in M^3 in the same way as the output of BCBT model. That's why we decide to divide the presentation of following M^3 implementation of data for considered auction models into four parts. First part concerns common data of aforementioned auction models for which we employ the data structure proposed in [4]. The second and third parts relate to the implementation of demand data concerning VPN specified in pipe and hose model, respectively. In fourth part we mention about representation of market clearing results.

In the following M^3 specification we present the snippets of XML files, that refer to two illustrative examples of bandwidth allocation problem: example E1 relating to VPN_PIPE auction model and example E2 relating to VPN_HOSE auction model. The market data for both this examples are presented in figures on page 183. Examples E1 and E2 regard the same telecommunication network consisting of 6 nodes and 6 pairs of directed links between them (see Fig. 1). In both examples at the supply side there is one seller associated with each link that is willing to sell 5 units of bandwidth at minimum unit price equal 1. In Fig. 1 the maximum volume and unit price of sell offers are denoted by the first and second value in parentheses, respectively. In the example E1 there is one buyer, whose buy offer is described in Fig. 2. The buyer specifies VPN traffic demands in the pipe model. The VPN has three endpoints: B, E and F. Demands between endpoints are depicted by dotted arrows with values denoting maximum demand volume. The buyer is willing to pay for the whole VPN at most 20 monetary units. In the example E2 there is also one buyer that submits offer illustrated in Fig. 3. The buy offer regards VPN that is

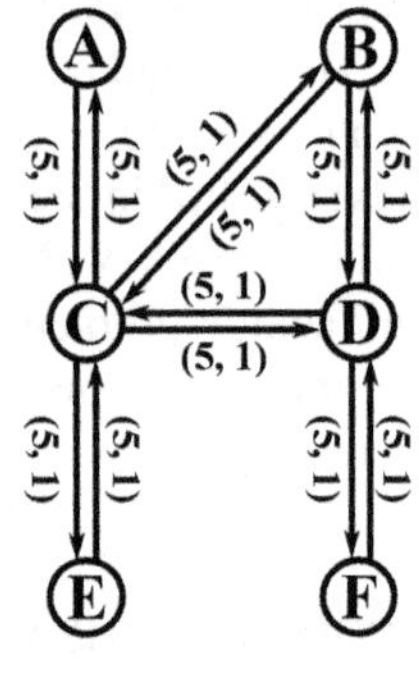

Fig. 1 Physical network and sell offers for both examples E1 and E2

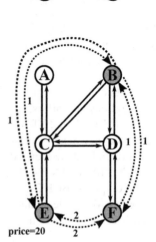

Fig. 2 Buy offer for example E1 specified in the pipe model

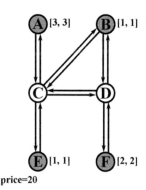

Fig. 3 Buy offer for example E2 specified in the hose model

defined in hose model. VPN consists of four endpoints: A, B, E and F. In Fig. 3 with each endpoint there are connected two values in brackets that denote the ingress and egress bandwidth for each endpoint, respectively. The buyer is willing to purchase a whole VPN at price not greater than 20 monetary units.

3.1 Physical Network and Supply Side Data

We begin from specifying the common part of data for VPN_PIPE and VPN_HOSE auction models, i.e. physical network and data associated with the supply side, such as an information about sellers, commodities for sale and sell offers. As it has been already mentioned, all this data can be represented in the same way, as it has been done in the case of BCBT model in [4]. However, in order to make the chapter self-contained, we present here the most important elements of this specification that are relevant to considered VPN auction models.

The network infrastructure can be modelled in M^3 as several acyclic graphs. One of this graphs is called basic network and it is the most detailed infrastructure model. Further graphs are called virtual networks. Each virtual network is an aggregation of a lower level virtual or basic network. In the case of considered VPN market, the basic network represents physical telecommunication network. We assume that nodes and arcs of physical network are instances of "PhysicalNode" and "PhysicalLink", respectively. "PhysicalNode" and "PhysicalLink" are kinds of nodes and arcs respectively that are defined in M^3 dictionary in the m3:networkKinds element.

Physical network with all its nodes and arcs is described in the m3:Network element, which is a child of the m3:networks element containing data about all networks. Each node and arc is represented by m3:node and m3:arc element, respectively. The m3:node (m3:arc) element has unique id of node (arc), child element m3:name containing the name of node (arc) and attribute dref referring to the node (arc) type. Additionally, the m3:arc element has the m3:predecessor element that has a reference to a predecessor of arc and the m3:successor element that has a reference to a successor of arc. A snippet of XML file containing data about physical network for examples E1 and E2 is presented on Listing 1.

Listing 1 Definition of the physical network in examples E1 and E2 (partial)

```
<m3:Network isDirected="true" id="SixNodesPhysicalNetwork">
   <m3:name>Physical telecommunication network</m3:name>
   <m3:node dref="PhysicalNode" id="A">
      <m3:name>Node A</m3:name>
   </m3:node>
   <!-- ... -->
   <m3:arc dref="PhysicalLink" id="link_A-C">
      <m3:name>Arc connecting nodes A and C</m3:name>
      <m3:predecessor ref="A"/>
      <m3:successor ref="C"/>
   </m3:arc>
   <!-- ... -->
</m3:Network>
```

M^3 model enables representing data of entities that participate in the VPN market. All types of market entities are specified in the m3:market- EntityKinds element of M^3 dictionary, in which we define type "link-seller" representing the sellers of link bandwidth. Information about particular sellers is stored in the m3:marketEntities element. Each market entity is specified in m3:MarketEntity that has unique id, child element m3:name containing its name, and attribute dref referring to the type of market entity. In the examples E1 and E2 the seller associated with link "link_A-C" can be described in a way shown on Listing 2.

Listing 2 Definition of the particular seller in examples E1 and E2

```
<m3:MarketEntity dref="link-seller" id="linkA-C_seller">
   <m3:name>Link A-C seller</m3:name>
</m3:MarketEntity>
```

Different kinds of commodities can be defined in M^3 dictionary in the commodityKinds element. In this element we specify type of commodity "link_bandwidth" that comprises all commodities related to the bandwidth of single link. In M^3 each commodity is associated with node or link and some period of time. Commodity of type "link_bandwidth" is available at the arc of the basic network. For illustrative purpose of examples E1 and E2 we assume that all commodities concerns one period of time, e.g. one month (January 2011) defined in M^3 calendar element and identified by "Jan2011". Commodities are defined in the m3:commodities element as its child elements m3:Commodity that have unique id and attribute dref referring to the type of commodity and child elements: m3:description that contains human readable description of commodity, m3:availableAt that has a reference to the link at which commodity is available and m3:CalendarScheduledCommodity that has a reference to time period with which this commodity is connected. A bandwidth of link "link_A-C" from examples E1 and E2 can be defined in M^3 as a commodity on Listing 3.

Listing 3 Definition of the commodity representing a bandwidth of link "link_A-C" in examples E1 and E2

```
<m3:Commodity dref="link_bandwidth" id="A-C_link_bandwidth">
   <m3:description>The bandwidth of link A-C available
                   in January 2011
   </m3:description>
   <m3:availableAt ref="link_A-C"/>
   <m3:CalendarScheduledCommodity ref="Jan2011"/>
</m3:Commodity>
```

Having defined data of physical network, sellers and commodities for sale, we are ready to specify sell offers. Sell offers in both VPN_PIPE and VPN_HOSE auction models concern single telecommunication links and can be defined as simple offers that are described by admissible range of commodity volumes and a unit price. All offers at the market are defined in the m3:offers element. Each offer is represented by the m3:Offer element that contains information about offered price, market entity that submits the offer, offer status, minimum and maximum volume of offer and commodity that the offer is related to. Returning to the examples E1 and E2, the offer submitted by seller "linkA-C_seller" on the commodity "A-C_link_bandwidth" can be specified in M^3, as shown on Listing 4.

Listing 4 Definition of the offer submitted by seller "linkA-C_seller" in examples E1 and E2

```
<m3:Offer offeredPrice="1" id="linkA-C_sell_offer">
   <m3:offeredBy ref="linkA-C_seller"/>
   <m3:offerStatus status="m3:offer-open">
      <m3:durationPeriod startTime="2011-01-01T00:00:00"
                         endTime="2011-02-01T00:00:00" />
   </m3:offerStatus>
   <m3:volumeRange minValue="0" maxValue="5" />
   <m3:ElementaryOffer>
      <m3:offeredCommodity ref="A-C_link_bandwidth"
                           shareFactor="1"/>
   </m3:ElementaryOffer>
</m3:Offer>
```

3.2 Demand Data Concerning VPN in the Pipe Model

To model the demand data concerning VPN specified in the pipe model we define another network – virtual network. In M^3 dictionary we add to the m3:networkKinds element a new type of node – "DemandVirtualNode" and new type of arc – "DemandVirtualArc" that are associated with that virtual network. The virtual network has the same number of nodes as physical network with each node aggregating one node of physical network. The relation of aggregation is specified in the m3:aggregates element included in the m3:node element defining particular node of virtual network. The virtual network is a complete graph, i.e. between each two nodes there is an arc that represents single demand. A part of the m3:Network element defining virtual network for example E1 is presented on Listing 5.

Listing 5 Definition of the virtual network in the example E1 (partial)

```
<m3:Network isDirected="true" id="SixNodesVirtualDemandNetwork">
   <m3:name>Virtual network representing demands</m3:name>
   <m3:VirtualNetwork aggregationType="demandTrade"/>
   <!-- ... -->
   <m3:node dref="DemandVirtualNode" id="virt_node_B">
      <m3:name>Virtual node B</m3:name>
      <m3:aggregates ref="node_B"/>
   </m3:node>
   <!-- ... -->
   <m3:arc dref="DemandVirtualArc" id="demand_B-F">
      <m3:name>Demand between virtual nodes B and F</m3:name>
      <m3:predecessor ref="virt_node_B"/>
      <m3:successor ref="virt_node_F"/>
   </m3:arc>
   <!-- ... -->
</m3:Network>
```

Data about VPN buyers can be defined analogously as data about the sellers of link bandwidth. In M^3 dictionary in the m3:marketEntityKinds element we specify a new type of market entity, namely "vpn_pipe-buyer", that represents the kind of entity that is willing to purchase a VPN specified in the pipe model. Information concerning the buyer from example E1 can be included in the m3:MarketEntity element, as shown on Listing 6.

Listing 6 Definition of the buyer in the example E1

```
<m3:MarketEntity dref="vpn_pipe-buyer" id="vpn_pipe_buyer1">
   <m3:name>Buyer 1 of VPN specified in pipe model </m3:name>
</m3:MarketEntity>
```

In the commodityKinds element we define a "demand_bandwidth" as a kind of commodity that concerns single demand. Commodities of type "demand_bandwidth" are related with arcs of virtual network. In the example E1, the demand for bandwidth between nodes B and F can be specified as the m3:Commodity element, as shown on Listing 7.

Listing 7 Definition of the commodity representing a demand for a bandwidth between nodes B and F in the example E1

```
<m3:Commodity dref="demand_bandwidth" id="dB-F">
   <m3:description>Demand for bandwidth between nodes B and F
                  required in January 2011
   </m3:description>
   <m3:availableAt ref="demand_B-F"/>
   <m3:CalendarScheduledCommodity ref="Jan2011"/>
</m3:Commodity>
```

Finally, we can represent buy offer concerning VPN specified in pipe model as a bundled offer. Bundled offer is a type of offer in M^3 that enables market entity to trade with bundles of commodities with fixed proportions of commodities defined in the offer. The m3:Offer element concerning bundled offer has the m3:BundledOffer element that contains references to all commodities that belong to the bundle with information about their proportions (attribute shareFactor). The buy offer from example E1 can be represented in M^3 in a way presented on Listing 8:

Listing 8 Definition of the buy offer in the example E1

```
<m3:Offer offeredPrice="-20" id="vpn_pipe_offer_1">
   <m3:offeredBy ref="vpn_pipe_buyer1"/>
   <m3:offerStatus status="m3:offer-open">
      <m3:durationPeriod startTime="2009-01-01T00:00:00"
                         endTime="2009-02-01T00:00:00"/>
   </m3:offerStatus>
   <m3:volumeRange minValue="0" maxValue="1"/>
   <m3:BundledOffer>
      <m3:offeredCommodity ref="dB-F" shareFactor="-1"/>
      <m3:offeredCommodity ref="dF-B" shareFactor="-1"/>
      <m3:offeredCommodity ref="dB-E" shareFactor="-1"/>
      <m3:offeredCommodity ref="dE-B" shareFactor="-1"/>
      <m3:offeredCommodity ref="dE-F" shareFactor="-2"/>
      <m3:offeredCommodity ref="dF-E" shareFactor="-2"/>
   </m3:BundledOffer>
</m3:Offer>
```

3.3   Demand Data Concerning VPN in the Hose Model

To model the demand data concerning VPN specified in the hose model, we define two virtual networks for each buy offer: VN1 and VN2. Virtual network VN1 consists of nodes each one aggregating single node of physical network corresponding to certain endpoint of VPN specified in the buy offer. Each node of this virtual network is of type "HoseVirtualNode" specified in M^3 dictionary in the m3:networkKinds element, as shown on Listing 9:

Note that for the node of "HoseVirtualNode" two M^3 parameters are defined: "ingress_bandwidth" and "egress_bandwidth" denoting ingress and egress bandwidth of the VPN endpoint, respectively. The virtual network VN2 contains only one node that aggregates all nodes of virtual network VN1 related to the same buy offer. This node is of type "VPNHoseAggNode" that is specified in M^3 dictionary in the m3:networkKinds element. The virtual networks VN1 and VN2 have no arcs.

Figure 4 shows the virtual network VN1 defined for the buy offer from the example E2. VPN specified in this buy offer consist of four endpoints represented by nodes A, B, E i F of physical network. Each of this nodes has corresponding node in virtual network VN1 related to this buy offer that contains information about ingress and egress bandwidth of related VPN endpoint. On Listing 10 we present a part of the m3:Network element defining virtual network VN1 for buy offer from the example E2.

Listing 9 Definition of the node type for a virtual network VN1 in the example E2 (a part of the m3:networkKinds element)

```
<m3:networkKinds>
    <m3:NetworkNodeKind id="HoseVirtualNode">
        <m3:name>Endpoint of VPN in the hose model</m3:name>
        <m3:typeParameter dref="ingress_bandwidth"/>
        <m3:typeParameter dref="egress_bandwidth"/>
    </m3:NetworkNodeKind>
    <m3:ParameterDefinition  dataType="xsd:decimal"
    id="ingress_bandwidth" required="true" unitOfMeasure="Mbps">
        <m3:name>Ingress bandwidth</m3:name>
    </m3:ParameterDefinition>
    <m3:ParameterDefinition  dataType="xsd:decimal"
    id="egress_bandwidth" required="true" unitOfMeasure="Mbps">
        <m3:name>Egress bandwidth</m3:name>
    </m3:ParameterDefinition>
    <!-- ... -->
</m3:networkKinds>
```

Listing 10 Definition of the virtual network VN1 in the example E2 (partial)

```
<m3:Network isDirected="true" id="VN1_ABEF">
    <m3:name>Virtual network VN1_ABEF</m3:name>
    <m3:VirtualNetwork aggregationType="hoseEndpoints"/>
    <m3:node dref="HoseVirtualNode" id="A1">
        <m3:name>Node aggregating endpoint A</m3:name>
        <m3:aggregates ref="A"/>
        <m3:parameter dref="ingress_bandwidth">3</m3:parameter>
        <m3:parameter dref="egress_bandwidth">3</m3:parameter>
    </m3:node>
    <!-- ... -->
</m3:Network>
```

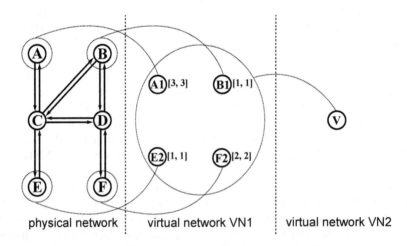

Fig. 4 Virtual networks used for modelling buy offer from example E2

Figure 4 also presents the virtual network VN2 for the buy offer from the example E2. This virtual network has only one node V that aggregates all nodes of virtual network VN1 related to the same buy offer. Node V is used to define the commodity that represents VPN service specified in the hose model. This commodity is of type "vpn_hose_service" defined in the commodityKinds element. The commodity of type "vpn_hose_service" is available at node of type "VPNHoseAggNode". In the M^3 model we can represent a VPN service required by the buyer from the example E2 as shown on Listing 11.

Listing 11 Definition of the commodity representing a VPN service in the example E2

```
<m3:Commodity dref="vpn_hose_service" id="VPN_ABEF_service">
   <m3:description>VPN service specified in the hose model
                  that is needed in January 2011
   </m3:description>
   <m3:availableAt ref="V"/>
   <m3:CalendarScheduledCommodity ref="Jan2011"/>
</m3:Commodity>
```

The buy offer concerning VPN specified in hose model can be now represented as a M^3 simple offer. Assuming that the buyer from the example E2 is defined in M^3 dictionary and its identifier is "vpn_hose_buyer1" the specification of his buy offer can be defined as on Listing 12.

Listing 12 Definition of the buy offer in the example E2

```
<m3:Offer offeredPrice="-20" id="buy_offer_vpn1">
   <m3:offeredBy ref="vpn_hose_buyer1"/>
   <m3:offerStatus status="m3:offer-open">
      <m3:durationPeriod startTime="2009-01-01T00:00:00"
                         endTime="2009-02-01T00:00:00"/>
   </m3:offerStatus>
   <m3:volumeRange minValue="0" maxValue="1"/>
   <m3:ElementaryOffer>
      <m3:offeredCommodity ref="VPN_ABEF_service"
                           shareFactor="-1"/>
   </m3:ElementaryOffer>
</m3:Offer>
```

3.4   Data Describing Market Clearing Results

Above we show how input data for VPN_PIPE and VPN_HOSE auction models can be defined in M^3. The output data resulting from clearing the market can also be easily specified in M^3. Figures on page 190 presents the realisation of buy offers determined by VPN_PIPE and VPN_HOSE auction models for examples E1 and E2, respectively. In both examples buy offers are fully accepted. Values at links denote the bandwidth allocated to buyer that provides required VPN service (links at which buyer does not obtain any bandwidth are not shown). Clearing price of the each link is set to the offered price specified in associated sell offer and the buyer

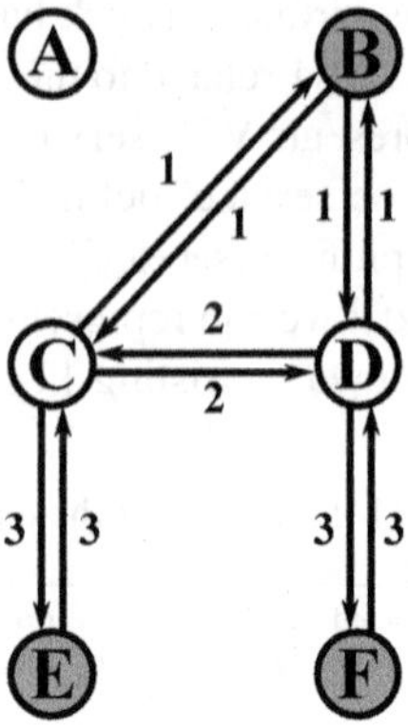

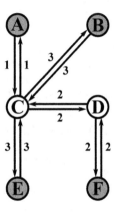

Fig. 5 Bundle of links realizing the buy offer from the example E1

Fig. 6 Bundle of links realizing the buy offer from the example E2

pays the sum of prices of links at which he obtains the bandwidth. The results of market clearing can be specified as the output offers represented in M^3 by the same m3:Offer element that is used for input sell and buy offers specification. For each input offer the corresponding output offer is created after clearing the market. Output offer corresponding to the sell offer is a simple offer referring to the commodity representing a bandwidth of link that is offered for sale in this sell offer. Note that in the case of both types of VPN specification (pipe and hose) the resulting allocation of bandwidth links realizing VPN is a bundle of links with appropriately set proportions of bandwidth at particular links. Thus, output offer corresponding to the buy offer related to VPN specified in the pipe or hose model can be represented as the bundled offer that concerns a bundle of commodities representing the bandwidth of relevant links with appropriate values of shareFactor attribute.

4 Conclusions

We have shown how M^3 model can be applied to describe a complex market data involved with trading VPN services. Two VPN representations were considered, namely pipe and hose model. In the case of pipe model, the bundled offer concept was used that enable us to define appropriate buy offer. In the case of hose model, two layers of virtual network were set that allow for defining relevant commodity and simple buy offer. We belive that M^3 model can be easily applied to many other segments of bandwidth market. It already has been used in the experimental environment for implementing different bandwidth auction models such as that presented in [2, 3, 5, 9].

References

1. Duffield, N.G., Goyal, P., Greenberg, A., Mishra, P., Ramakrishnan, K.K., van der Merwe, J.E.: A flexible model for resource management in virtual private networks. SIG-COMM Comput. Commun. Rev. 29, 95–108 (1999), doi:10.1145/316194.316209
2. Jain, R., Varaiya, P.: Efficient Market Mechanisms for Network Resource Allocation. In: Proc. 44th IEEE Conf. on Decision Control, pp. 1056–1061. IEEE (2005)
3. Jain, R., Walrand, J.: An Efficient Mechanism for Network Bandwidth Auction. In: Network Operations and Management Symposium Workshops, pp. 227–234. IEEE (2008)
4. Kacprzak, P., Kaleta, M., Pałka, P., Smolira, K., Toczyłowski, E., Traczyk, T.: Application of open multi-commodity market data model on the communication bandwidth market. Journal of Telecommunication and Information Technology 4, 45–50 (2007)
5. Kołtyś, K., Pałka, P., Toczyłowski, E., Żółtowska, I.: Multicommodity auction model for indivisible network resource allocation. Journal of Telecommunication and Information Technology 4, 60–66 (2008)
6. Kołtyś, K., Pieńkosz, K., Toczyłowski, E., Żółtowska, I.: Auction model for bandwidth trading with possibility of purchasing VPN. Przegląd Telekomunikacyjny 8 9, 1183–1189 (2009) (in Polish)
7. Kołtyś, K., Pieńkosz, K.: Auction mechanisms for trading virtual private networks in the hose model. Przegląd Telekomunikacyjny 8-9, 1003–1010 (2010)
8. Kołtyś, K., Pieńkosz, K., Toczyłowski, E.: A Flexible Auction Model for Virtual Private Networks. In: Domingo-Pascual, J., Manzoni, P., Palazzo, S., Pont, A., Scoglio, C. (eds.) NETWORKING 2011, Part II. LNCS, vol. 6641, pp. 97–108. Springer, Heidelberg (2011), doi:10.1007/978-3-642-20798-3_8
9. Pałka, P., Kołtyś, K., Toczyłowski, E., Żółtowska, I.: Model for Balancing Aggregated Communication Bandwidth Resources. Journal of Telecommunication and Information Technology 3, 43–49 (2009)
10. Stańczuk, W., Lubacz, J., Toczyłowski, E.: Trading links and paths on a communication bandwidth markets. J. Univers. Comput. Sci. 14, 642–652 (2008)